ST(P) MATHEMATICS 5C

Teacher's Notes and Answers

L. Bostock, B.Sc.

S. Chandler, B.Sc.

A. Shepherd, B.Sc.
Head of Mathematics, Redland High School for Girls

E. Smith, M.Sc.
Head of Mathematics, Tredegar Comprehensive School

Stanley Thornes (Publishers) Ltd

INTRODUCTION

5C covers the work required for the written papers of the foundation level of GCSE. This book is intended as a two year course, leading up to the examination at sixteen plus. It can also be used, just in the last year, for those pupils who were aiming at intermediate level but who, it has now become clear, should be entered for foundation level.

No one syllabus requires all the topics in this book, so a careful check of the particular syllabus being studied is necessary.

The topics have been arranged in separate chapters so that pupils can easily find topics for revision purposes. The order of the chapters is not intended to imply a teaching order or a level of difficulty. Many sections are self-contained and can be covered at any time. Chapter 33, for example, starts with some easy work on bearings which can be studied early on in the course and then revised before doing the more complicated work involved in scale drawings at the end of the chapter.

It is also possible, and probably desirable at this level, to dip into some of the basic skills chapters for revision before covering applications. For example, parts of Chapter 5 on decimals can be used as a reminder before working the first part of Chapter 18, Earning Money. Chapter 17 on percentages can be used in a similar way before working the remainder of chapter 18. It is sensible to refer back to the chapters on basic processes every time that they need to be used in applications.

Many pupils find it difficult to appreciate the need for basic numeracy and, in particular, the need to understand percentages. One way to encourage effort is to point out that people who go through life using very little mathematics are at a considerable disadvantage. They have to trust 'experts' and will not be able to check when they feel that they may have been 'done'. For example, has the salesperson in the carpet shop given me the right estimate for the amount of carpet I need for the hall or am I being sold a bit more than I really need; has the travel agent done his sums right on the holiday I want to book; is the union official right in saying that a 10% rise is better than a flat rate rise of £20 a week, ...? In other words, emphasise that they are learning skills for living in the adult world.

There are further notes on each chapter giving ideas on relevance and suggestions for extensions and projects.

We have not included mixed revision exercises because past papers are a better source at this stage.

CHAPTER 1   Whole Number Arithmetic

This chapter gives basic practice in dealing with whole numbers. It also contains some work on estimating. Ask pupils what they would do if they bought two items costing 37 p and 25 p respectively and they were asked to pay 74 p. If they know that $37 + 25$ is roughly 60 then they can say that 74 p is wrong, and then if they can add 37 and 25 without a calculator they will also know what the correct amount should be. You could ask them for other situations in which the ability to estimate, and to add without a calculator, is an advantage.

Exercise 1a   p. 1

1. 38
2. 232
3. 704
4. 2423
5. 22 600
6. Seventy-eight
7. Four hundred and thirty-seven
8. Six thousand, two hundred and eighty
9. Thirty-two
10. Five hundred and eight
11. Two thousand and twenty
12. 1892
13. a) One thousand nine hundred and thirty-four
    b) Nineteen thirty-four
14. 700
15. 3, 16, 19, 56, 61
16. 243
17. a) Nine is greater than three.
    b) Seven is less than eight.
18. a) $12 > 7$ b) $9 < 14$
19. 14, 17, 41, 47, 71, 74

Exercise 1b   p. 3

Encourage pupils to provide examples like these from newspapers.

1. a) £12 000   c) £20 000
   b) £7000   d) £8500
2. a) £3 000 000   b) £2 500 000
3. £5 000 000 000
4. £1 500 000
5. a) 10 000 000 000
   b) 150 000 000
   c) 4 500 000   d) 30 000

Exercise 1c   p. 4

1. 44
2. 17
3. 180
4. 62
5. 693
6. 24
7. 8925
8. 124
9. 22
10. 18
11. 22
12. 25
13. 30
14. 129
15. 66
16. 39
17. 6
18. 13
19. 20
20. 18
21. 5
22. 20
23. 21
24. 18
25. 9
26. 7

Exercise 1d   p. 5

1. 4
2. 5
3. 5
4. 4
5. 7
6. 8
7. 7
8. 12
9. 5
10. 102
11. 1984
12.

+	4	5	6
1	5	6	7
2	6	7	8
3	7	8	9

13. a)

−	3	4	5
7	4	3	2
8	5	4	3
9	6	5	4

b)

−	2	3	5
6	4	3	1
8	6	5	4
10	8	7	6

14. a)
$$\begin{array}{r} 168 \\ +351 \\ \hline 519 \end{array}$$

b)
$$\begin{array}{r} 48 \\ -25 \\ \hline 23 \end{array}$$

15. 1976

Exercise 1e   p. 6

1. 493
2. 182
3. 747
4. 1008
5. 672
6. 7056
7. 1728
8. 2592
9. 384
10. 798
11. 165

Exercise 1f   p. 7

1. 24
2. 72
3. 54
4. 63

5. 84
6. 96
7. 85
8. 90

9. 56
10. 60
11. 90
12. 42

Exercise 1g   p. 8

1. 700
2. 600
3. 620
4. 32 000
5. 9000
6. 900

7. 1000
8. 8000
9. 6000
10. 800
11. 9000
12. 1200

13. 1200
14. 1200
15. 240 000
16. 16 000

Exercise 1h   p. 8

1. 7
2. 6
3. 12
4. 30
5. 8
6. 90
7. 30
8. 20
9. 30

10. 40
11. 2
12. 40
13. 24
14. 48
15. 12
16. 18
17. 24
18. 64

19. 27
20. 16
21. 2
22. 3
23. 1
24. 1
25. 2
26. 6 packets, 10 p

Exercise 1i   p. 10

Encourage pupils to provide data supplied by the media.

1. 400, 800, 2400, 300, 7600, 100
2. 20, 40, 70, 320, 1370, 10
3. 4000, 6000, 1000, 9000, 41 000
4. a) 400 injured   b) 420

Exercise 1j   p. 11

1. 798 $(40 \times 20 = 800)$
2. 1536 $(30 \times 50 = 1500)$
3. 27 540 $(100 \times 300 = 30 000)$
4. 504 $(60 \times 10 = 600)$
5. 847 $(80 \times 10 = 800)$
6. 522 $(20 \times 30 = 600)$

7. 6572 $(200 \times 30 = 6000)$
8. 12 320 $(400 \times 30 = 12 000)$
9. 2 004 402 $(700 \times 3000 = 2 100 000)$
10. 10 962 $(90 \times 100 = 9000)$
11. 1102 $(20 \times 60 = 1200)$
12. 4590 $(300 \times 20 = 6000)$

Exercise 1k   p. 11

1. 4
2. 6200
3. 50 205

4. a) i) 7   ii) 3   iii) 9   b) WAGER
c) e.g. RAG (4), WAR (6), WAGE (8),
RAGE (5), AGE (4), GEARS (5)

Exercise 1l   p. 12

1. 42
2. a) £2.73   b) 12 p
3. 8 boxes; 4 eggs are left.

4.
$$\begin{array}{r} 336 \\ + 926 \\ \hline 1262 \end{array} \qquad \begin{array}{r} 6338 \\ - 1723 \\ \hline 4615 \end{array}$$

5. 20 p, 20 p, 2 p, 2 p, 2 p, 1 p or
10 p, 10 p, 10 p, 10 p, 5 p, 2 p or
20 p, 10 p, 10 p, 5 p, 1 p, 1 p

Exercise 1m   p. 12

1. a)

×	1	2	3
2	2	4	6
4	4	8	12
6	6	12	18

b)

×	0	1	2
3	0	3	6
5	0	5	10
7	0	7	14

2. 11
3. 299
4. a) 2
   b) 4 p
5. a)   i) 252   ii) 261   iii) 248
   b) 248, 252, 261

Exercise 1n   p. 13

1. 26
2. 7
3. 16
4. 20
5. 40
6. 3
7. 18
8. 3
9. 21
10. 12
11. 33
12. 21
13. 4
14. 17
15. 19
16. 0
17. 40
18. 14
19. 8
20. 22
21. 2
22. 25
23. 2
24. 1
25. 24
26. 2
27. 12
28. 1
29. a) 3
    b) 2

Exercise 1p   p. 14

In question 17(b), point out that calculators vary in the way that they handle a mixture of operations.

1. 28
2. 20
3. 12
4. 13
5. 3
6. 2
7. 14
8. 19
9. 27
10. 12
11. 45
12. 42
13. 18
14. 17
15. 1
16. 13
17. a) 23
    b) 20
18. a) 18
    b) Anthony might get 18 or 30

Exercise 1q   p. 15

1. a) 14   b) 7   c) 6   d) 8
2. 8 p
3. 192
4. a) 6   b) 2
5. a) 2720   b) 15 980
6. a) 594   b) 54

CHAPTER 2   Numbers and Number Patterns

This chapter is also on whole numbers, dealing this time with their nature and make-up in terms of factors and square roots and the patterns that can be formed with them.

Exercise 2a   p. 16

1. 3, 12, 15, 36
2. Any three from 12, 14, 16, 18, 20
3. 7, 13, 15, 59
4. 4, 40, 400
5. a) Yes   b) No
6. Yes

Exercise 2b   p. 17

1. 1, 2, 3, 4, 6, 12
2. 1, 2, 4, 5, 10, 20
3. 1, 3, 9, 27
4. 1, 2, 4, 8
5. 1, 2, 4, 8, 16
6. 1, 2, 3, 5, 6, 10, 15, 30
7. 1, 2, 3, 4, 5, 6, 10, 12, 15, 20, 30, 60
8. 1, 2, 3, 6, 9, 18, 27, 54
9. No
10. Yes
11. Yes
12. 1, 7
13. 2, 4, 6, 8, 12, 16, 24, 48

Exercise 2c   p. 18

1. 2, 3, 11

2. 2, 3, 5, 7

3. Factors are 1, 2, 3, 5, 6, 10, 15, 30
The prime factors are 2, 3, 5

4. Yes. No other even numbers are
prime.

5. 2, 3

6. a) 2, 4, 8, 12   c) 4, 8, 12
b) 2, 3, 7   d) 2, 3, 4, 8, 12

7. a) 3, 5, 7, 9   c) 3, 9, 12
b) 3, 5, 7   d) 3, 7

8. 9

9. a) 27   b) 101

Exercise 2d   p. 19

1. a) 25   b) 121   c) 6724

2. a) 5   b) 7   c) 10

3. a) 8   b) 64   c) 125

4. 6

5. 36

6. a) 49   b) 100   c) 64

7. a) 2   b) 4   c) 9

8. 8

Exercise 2e   p. 20

1. 9, 11, 13, . . .

2. 15, 19, 23, . . .

3. 14, 12, 10, . . .

4. 54, 162, 486, . . .

5. 8, 4, 2, . . .

6. 81, 243, 729, . . .

7. 25, 36, 49, . . .

8. 3000, 30 000, 300 000, . . .

9. 48, 96, 192, . . .

10. 1, 4, 9, 16,

```
•  •  •  •  •

•  •  •  •  •

•  •  •  •  •

•  •  •  •  •

•  •  •  •  •
```
25, 36, . . .

11. 1, 3, 6, 10,
```
         •

       •   •

     •   •   •

   •   •   •   •

 •   •   •   •   •
```
15; 21; . . .

12. 1, 3, 5, 7,
```
   •                •
   •                •
   •                •
   •                •
   • • • •      • • • • •
```
9; 11; add 2

13. $(3+4) \times 5 = 35$
$(4+5) \times 6 = 54$
$(5+6) \times 7 = 77$
$(6+7) \times 8 = 104$

Question 13 can be extended to investigate the sequence formed from the differences between the terms
  i.e.   11, 14, 19, 23, 27 . . .
  then   4, 4, 4, 4, . . .

Some number patterns similar to that in question 13 are:

a) $1 + 2 + 3 =$
$2 + 3 + 4 =$
$3 + 4 + 5 =$
  . . .

b) $2 \times 1 + 2 =$
$2 \times 2 + 3 =$
$2 \times 3 + 4 =$
  . . .

c) $1 + (2 \times 3) =$
$2 + (3 \times 4) =$
$3 + (4 \times 5) =$
  . . .

Exercise 2f   p. 21

1. 80, 160, . . .

2. 25, 36
a) i) 1, 9, 25   ii) 4, 16, 36
b) e.g. 16 can be represented by
```
  •  •  •  •

  •  •  •  •

  •  •  •  •

  •  •  •  •
```

3. a) $4 \times 5, 5 \times 6, \ldots$
b) 11, 13, . . .

4. $1 + 2 + 3 + 4 = 10$
$1 + 2 + 3 + 4 + 5 = 15$

5. a) 11, 9, . . .   b) 17, 21, . . .

6. $1 + 3 + 5 + 7 + 9 = 25 = 5^2$
$1 + 3 + 5 + 7 + 9 + 11 = 36 = 6^2$

7. $(6 \times 7) - 5 = 36 + 1$

8. $2 + 4 + 6 + 8 + 10 = 30 = 5 \times 6$
$2 + 4 + 6 + 8 + 10 + 12 = 42$
$\qquad\qquad\qquad\qquad = 6 \times 7$

CHAPTER 3   Fractions

Fractions are used a great deal in everyday life to describe quantities such as money and areas. The need to understand fractions could be introduced by considering a situation like this:

I want to lay tiles on the kitchen floor. When I go to the shop there is one complete packet of tiles and one packet that has been opened. My floor has an area of 3 square metres and one packet covers 2 square metres. The assistant says that the full packet will do $\frac{2}{3}$ of the floor and the opened packet has enough to cover $\frac{1}{2}$ the floor, so the two together will be plenty to cover the whole floor. Is he right ? To answer that you need to know what $\frac{2}{3}$ etc. means, what $\frac{2}{3}$ of something looks like and what $\frac{2}{3}$ plus $\frac{1}{2}$ is.

Exercise 3a   p. 22

When this exercise has been completed, the pupils can draw a rectangular floor, 3 squares by 1 square and use shading to answer the question posed at the beginning of this chapter.

1. $\frac{1}{4}$

2. $\frac{3}{8}$

3. $\frac{7}{10}$

4. $\frac{5}{8}$

5. $\frac{3}{16}$

6. $\frac{1}{2}$

7.

8.

9.

Exercise 3b   p. 24

1. a)

b) Yes

c) e.g. $\frac{4}{8}, \frac{5}{10}$

2. a)

b) No

3. No

4. 5

5. a) $\frac{1}{2}$ f) $\frac{9}{10}$

b) $\frac{2}{5}$ g) $\frac{2}{3}$

c) $\frac{2}{3}$ h) $\frac{3}{4}$

d) $\frac{1}{3}$ i) $\frac{7}{10}$

e) $\frac{1}{4}$

Exercise 3c   p. 25

1. e.g. $\frac{2}{6}, \frac{3}{9}$

2. a) $\frac{6}{10}$ b) $\frac{9}{15}$

3. $\frac{4}{8}$

4. 6 ninths

5. e.g. $\frac{6}{8}, \frac{12}{16}, \frac{9}{12}$

6. e.g. $\frac{1}{2}, \frac{2}{4}, \frac{4}{8}, \frac{3}{6}$

Exercise 3d   p. 25

1. 4

2. 2

3. 1

4. 3

5. 6

6. 1

7. 5

8. 1

9. $2\frac{1}{2}$

10. $1\frac{1}{3}$

11. $1\frac{3}{4}$

12. $4\frac{1}{2}$

13. $1\frac{2}{3}$

14. $1\frac{1}{2}$

15. $1\frac{1}{2}$

16. $2\frac{2}{3}$

Exercise 3e   p. 26

1. £3

2. 3 p

3. 48 p

4. 4 cm

5. 2 kg

6. 12 cm

7. £4

8. £9

9. 6

10. 7 m

11. £10

12. 18 boys, $\frac{1}{4}$ are girls

13. 10

14. £6

Exercise 3f p. 27

1. $\frac{1}{3}$
2. $\frac{11}{12}$
3. $\frac{7}{9}$
4. $\frac{5}{9}$
5. $\frac{1}{4}$
6. $\frac{7}{10}$

7. $\frac{7}{10}$
8. $\frac{4}{5}$
9. $\frac{2}{3}$
10. $\frac{1}{2}$
11. $\frac{4}{5}$
12. $\frac{1}{2}$

13. $\frac{1}{4}$
14. $\frac{2}{3}$
15. $\frac{1}{2}$
16. $\frac{1}{5}$

Exercise 3g p. 28

1. $\frac{1}{6}$
2. $\frac{1}{2}$
3. $\frac{3}{8}$
4. $\frac{5}{8}$

5. $\frac{3}{4}$
6. $\frac{5}{9}$
7. $\frac{2}{9}$
8. $\frac{2}{3}$

9. $\frac{1}{4}$
10. $\frac{1}{3}$
11. $\frac{1}{2}$
12. $\frac{3}{4}$

Exercise 3h p. 29

1. $\frac{1}{2}$
2. $\frac{2}{5}$
3. $\frac{2}{5}$

4. $\frac{1}{5}$
5. $\frac{1}{5}$
6. $\frac{1}{4}$

7. $\frac{11}{30}$
8. No. He had painted $\frac{3}{10}$ of the fence.

Exercise 3i p. 29

1. £2
2.

3. $\frac{4}{9}$
4. 5 p
5. $\frac{9}{12} = \frac{3}{4}$

6. $\frac{6}{7}$
7. $\frac{10}{12}$
8. $\frac{7}{10}$

Exercise 3j p. 30

1. $\frac{8}{10}$
2. 15 p
3. $\frac{5}{7}$

4.

5. $\frac{9}{12}$ and $\frac{3}{4}$

6. $\frac{3}{5}$
7. $\frac{8}{14} = \frac{4}{7}$
8. £8

CHAPTER 4 Time

Most pupils will be able to see the relevance of this topic but will probably think that they know all about it. It is surprising, though, how many people have difficulty in reading timetables and working out the number of Fridays, say, between two given dates.

Some of the exercises can be used for aural work.

Exercise 4a p. 31

1. January, February, March, April, May, June, July, August, September, October, November, December
2. June

3. February
4. June
5. October
6. a) June b) November c) March

7. a) February b) August c) October
8. No (June has 30 days.)
9. 7
10. 92

Exercise 4b   p. 32

The calendar for another year can be used for many of the questions in this exercise to give extra practice. The calendar for the current year could be used for questions on school-related problems, e.g. 'How many Mondays are there in this term ?' 'What about other days of the week ?' 'Does this mean that some subjects get slightly less teaching time than others ?' 'Does it even out over the whole school year ?, etc.

1. No
2. Thursday, 24 March
3. Monday, 4 September
4. 29 October, 21 October

5. a) 5 June   b) 31 May
6. 15
7. 18 June
8. 62

9. 47
10. a) 15   b) 15
11. a) 26 December   b) 15 August
     c) 10 June

Exercise 4c   p. 33

1. a) 15 min   b) 20 min
     c) 27 min   d) 27 min
2. a) 8.30 a.m.   b) 2.50 p.m.
     c) 7.10 a.m.   d) 8.35 a.m.
3. a) 10.50 a.m.   b) 5.18 p.m.
     c) 10.16 p.m.
     d) 11.02 p.m. (the previous day)

4. a) 55 min   b) 25 min
     c) 28 min   d) 45 min or $\frac{3}{4}$ h
5. 42 min
6. 3 h 25 min
7. 8.36 a.m.
8. a) 120   b) 300
     c) 150

9. a) 24   b) 72
     c) 168
10. a) 4   b) $1\frac{1}{2}$
11. a) 600   b) 1800
     c) 3600
12. a) 11 h 6 min   b) 9 h 45 min
     c) 13 h 32 min

Exercise 4d   p. 34

Most of the questions in qu. 1 can be used aurally for your own school timetable. Further aural work can be based on qus. 2 and 3 using recent details of TV and radio programmes taken from a newspaper.

1. a) i) 3   ii) 2   e) No
     b) 3 hr 35 min   f) 4 h 50 min
     c) 2 h 5 min   g) 1 h 30 min
     d) 1 h 5 min

2. a) 50 min
     b) 1 h 25 min
     c) Tuesday

3. a) 1 h 15 min
     b) 1 h 30 min
     c) 17 May
     d) 8 May

Exercise 4e   p. 36

The main problem that adults find with the 24-hour clock is the equivalence between p.m. times and 24-hour times. Plenty of aural work is needed and the timetables in exercises 4f, 4g and 4h can be used for this.

1. 0500
2. 2230
3. 2015
4. 1800
5. 0415
6. 1925
7. 1130
8. 0613
9. 1813
10. 0940

11. 1515
12. 2235
13. 7.30 a.m.
14. 2.00 p.m.
15. 3.40 p.m.
16. 10.15 a.m.
17. 11.14 p.m.
18. 0.10 a.m.
19. 8.19 a.m.
20. 9.32 p.m.

21. 10.39 a.m.
22. 5.20 a.m.
23. 1.08 p.m.
24. 6.15 p.m.
25. a) 5 h 34 min
     b) 5 h 7 min
     c) 10 h 10 min
26. 13 h 45 min
27. 1.32 p.m.
28. 6.30 a.m.

Exercise 4f   p. 37

This and the next two exercises contain extracts from timetables. Pupils should be able to see and use complete bus, coach and rail timetables: use local information wherever possible. These can also be used for projects such as planning an outing. For example they could be asked to work out an itinerary for a day trip to a place about 20 miles away, within certain constraints if desired, such as having to be back at a given time, or having to spend a given time at their destination.

1. 2 h 20 min, 2 h 15 min; no
2. 19 min, 14 min
3. a) 1 h 11 min   b) 50 min
4. a) 52 min   b) 14 min
5. a) Sidbury Cross and
        Martindale
     b) Martindale and Hardings
        Farm

6. Manley   1235
     Hardings Farm   1311
     Martindale arr   1325
             dep   1344
     Sidbury Cross   1436
     Lexton   1455
7. 1 h 35 min

Exercise 4g   p. 38

1. 6
2. a) 2   b) 4
3. The 1840, 59 minutes

4. 1 h 53 min
5. 1600
6. 1701, change at Bristol Parkway

7. a) 1 h 19 min   b) 50 min

Exercise 4h   p. 39

1. 7
2. 4
3. 1356, 6 min

4. a) 27 min   b) 35 min
5. a) 14 min   b) 26 min
6. 1305

7. 6
8. 11
9. 4

Exercise 4i   p. 40

1. 1800
2. 52 min
3. 20 May

4. No
5. 12 days

Exercise 4j   p. 40

1. 0800
2. 1 h 43 min

3. 13 November
4. 10 June

5. 9.37 p.m.
6. 15.11

CHAPTER 5   Decimals

The basic meaning of decimals is revised in the first two exercises with emphasis on place values. Students find the idea of place value difficult to grasp and it is worth spending plenty of time on questions such as 'What does the 3 in 2.36 mean ?' The next few exercises rely on understanding place value.

Exercise 5a   p. 42

1. 0.4
2. 0.03
3. 0.7
4. $\frac{7}{10}$ (or $\frac{1}{5}$)

5. $\frac{4}{100}$ (or $\frac{1}{25}$)
6. 2 units, 7 tenths, 1 hundredth
7. 4 hundredths or $\frac{4}{100}$
8. 3 tenths or $\frac{3}{10}$

9. a) 3 hundred or 300
     b) 3 units
     c) 3 tenths or $\frac{3}{10}$
10. a) 6   b) 2

Exercise 5b   p. 43

1. 23.14
2. 412.6
3. 3.27
4. 4.06

5. 50.2
6. a) 46.2
     b) 2 tenths or $\frac{2}{10}$
     c) 4 tens, or 40

7. $30 - 6 = 24$
8. $2 - 0.2 = 1.8$

Exercise 5c   p. 44

1. 0.75
2. 0.8
3. 0.375
4. 0.56
5. 0.12
6. 3.2

7. 8.5
8. 3.75
9. 2.4
10. 10.25
11. 6.5
12. 1.75

13. 5.3
14. 11.25
15. 7.09
16. 18.52

Exercise 5d   p. 45

1. 0.25	**10.** 0.4	**19.** 0.06, 0.23, 0.5
2. 0.6	**11.** 1.2	**20.** 0.06, 0.16, 0.6
3. 0.62	**12.** 0.492	**21.** 2.03, 2.3, 2.33
4. 0.87	**13.** 0.62	**22.** 1.08, 1.71, 1.73
5. 0.9	**14.** 6.45	**23.** 0.06, 0.6, 6
6. 0.76	**15.** 3.2	**24.** 0.625, 0.655, 0.713
7. 0.57	**16.** 0.7	**25.** 33
8. 8.08	**17.** 0.34	**26.** a) 0.8 c) $\frac{8}{10}$
9. 0.31	**18.** 0.1	b) 0.7 d) The first

Exercise 5e   p. 47

1. 16.3	**15.** 4.4	**29.** 0.08
2. 1.7	**16.** 0.1	**30.** 2.70
3. 73.6	**17.** 4.73	**31.** 22.58
4. 4.9	**18.** 3.09	**32.** 0.03
5. 19.4	**19.** 10.62	**33.** 32
6. 0.3	**20.** 0.87	**34.** 2
7. 5.9	**21.** 3.01	**35.** 17
8. 22.0	**22.** 7.40	**36.** 14
9. 13.3	**23.** 16.24	**37.** 7
10. 1.6	**24.** 1.97	**38.** 18
11. 3.1	**25.** 3.29	**39.** 111
12. 0.7	**26.** 4.02	**40.** 1
13. 32.4	**27.** 0.44	
14. 6.9	**28.** 0.06	

Exercise 5f   p. 48

1. a) 20   b) 18
c) 17.6   d) 17.63
2. a) 3   b) 2.7
c) 2.72
3. 0.857142 . . .
0.86 correct to nearest hundredth

4. 1.181818 . . .
1.2 correct to nearest tenth
5. 0.666666 . . .
0.67 correct to nearest hundredth
6. 1.555555 . . .
1.6 correct to nearest tenth

7. 5.666666 . . .
6 correct to nearest whole number

Exercise 5g   p. 49

Simple addition and subtraction should be done without a calculator.

1. 10.8	**12.** 18.07	**18.** 3.⬚2⬚ 6
2. 3.48	**13.** 14.31	1. 5 1
3. 2.31	**14.** 28.283	4. 7 7
4. 8.8	**15.** 12.1	
5. 2	**16.** 6.4	**19.** 2 .3⬚5⬚
6. 0.05	**17.** 5. 8 1	+⬚3⬚.1 6
7. 11.5	−3.⬚5⬚ 0	5 .5 1
8. 7.2	2. 3 1	**20.** 5 .⬚9⬚ 2
9. 12.9		+⬚2⬚. 2 ⬚3⬚
10. 1.1		8 . 1 5
11. 4.9		

Exercise 5h   p. 50 —————————————————————————————

In general, multiplication and division should be done with a calculator, but it is sensible to give some practice in multiplying a decimal by a whole number without a calculator. Use straightforward numbers such as 3×1.4, 2×1.8, 1.2×4, etc.

1. 9.36	**9.** 2.3	**17.** 0.09
2. 140	**10.** 9.1	**18.** 10.83
3. 14.4	**11.** 3.1	**19.** 4.37
4. 1.28	**12.** 5.3	**20.** 1.48
5. 3.46	**13.** 0.6	**21.** 2.67
6. 5	**14.** 10.4	**22.** 0.91
7. 3.84	**15.** 3.1	**23.** 0.44
8. 27.18	**16.** 0.4	**24.** 0.54

Exercise 5i   p. 51 —————————————————————————————

1. 9.61	**9.** 1.64	**17.** 1
2. 0.0036	**10.** 3.91	**18.** 0.072
3. 0.64	**11.** 4.36	**19.** 1.44
4. 7.29	**12.** 2.65	**20.** 0.56
5. 228.01	**13.** 0.77	**21.** 0.775
6. 53.29	**14.** 1.73	**22.** 7.84
7. 3.61	**15.** 3.32	**23.** 0.8
8. 0.04	**16.** 2.76	**24.** 35

Exercise 5j   p. 52 —————————————————————————————

1. 21.11	**8.** 9.2	**15.** 0.192
2. 4.816	**9.** 14.2	**16.** 465
3. 2.08	**10.** 91.5	**17.** 0.3844
4. 120.52	**11.** 681.21	**18.** 30.6
5. 43	**12.** 10.3	**19.** 11.3
6. 1.8	**13.** 6.59	**20.** 0.41
7. 368.64	**14.** 5.52	

Exercise 5k   p. 52 —————————————————————————————

1. 12 kg	**5.** 24 m	**7.** a) 0.2
2. 85 p	**6.** a) 0.75	b) 3.2
3. 38 cm	b) 1.75	c) 230.4
4. 40 p	c) £ 147	**8.** 459

Exercise 5l   p. 53 —————————————————————————————

1. 3.6 m	**3.** 6.4 kg	**5.** £ 3.30
2. 11.1 m	**4.** 126.4 cm	

Exercise 5m   p. 53 —————————————————————————————

Before working this exercise it is worth discussing what makes a sensible estimate and what is a ridiculous answer. For example:

David is driving in France and sees a speed limit sign of 100 km/h. Jenny uses her calculator and tells David to keep his speed down to 62.5 m.p.h. Peter roughly converts in his head and tells David that if he drives at about 50 m.p.h. he will be under the speed limit. Neither piece of advice is very sensible – why ?

1. $80 \div 8$; 10 m	**3.** $90 \div 10$; 9	**5.** $5 - 2$; 3 m
2. 20×3; 60 cm	**4.** $5 + 8$; 13 m	

Exercise 5n   p. 54

1. $\frac{7}{10}$ **4.** 4.8   **7.** 6 hundredths
2. 0.625   **5.** 10.24   **8.** 0.17
3. 8   **6.** 4

Exercise 5p   p. 54

1. 0.07, 0.59, 0.6   **4.** 7.7   **7.** 2 tenths
2. $\frac{9}{100}$ **5.** 0.02   **8.** 5.284
3. 1.7   **6.** 4.84

CHAPTER 6   Measuring Instruments

Link this work closely with estimating: motorists, for example, often need to estimate the amount of petrol in their tank and the distance that they can travel on it (running out of petrol can be very inconvenient). Pupils should have the opportunity to see and usc several dials and measuring scales, e.g. spring balance, kitchen scales, measuring tapes, ammeter, thermometer.

Exercise 6a   p. 55

1. a) 6 gallons   c) 3 gallons   **4.** a) 1 cm   b) 12 cm   **9.** a) $-16\,°C$ b) $4\,°C$
   b) 9 gallons   d) $7\frac{1}{2}$ gallons   **5.** a) 6 cm   b) 25 mm   c) $-13\,°C$
2. a) 50 m.p.h., 80 km/h   c) 60 mm   **10.** a) 70 ml   b) 20 ml
   b) 80 m.p.h., 128 km/h   **6.** a) 76 mm   b) 3.8 cm   c) 108 ml
   c) 94 m.p.h., 150 km/h   **7.** a) i) 28 °C   ii) 82 °F   **11.** a) 42   b) £ 17.85
   d) 56 m.p.h., 90 km/h   b) i) 59 °C   ii) 72 °F   c) 42.5 p
3. a) 50 m.p.h., 80 km/h   c) 20 °C   **12.** a) i) $2\frac{3}{4}$ inches   ii) 7 cm
   b) 75 m.p.h., 120 km/h   d) i) 27 °C   ii) 0 °C   b) 5.1 cm   c) $3\frac{1}{2}$ inches
   c) 100 m.p.h., 160 km/h   **8.** a) 6.5 cm   b) 4.9 cm

CHAPTER 7   Following Instructions

Exercise 7a   p. 61

The pupils can be asked to produce examples of real-life instructions that need numbers fed into them: cooking instructions on ready-packed chickens etc. are a good source.

1. a) 2 h 20 min   b) 2 h 50 min   **3.** a) £ 190   b) £ 140   **5.** a) £ 15   b) £ 17
2. a) 10 cm²   b) 12.6 cm²   **4.** a) 20 km   b) 37.5 miles   **6.** a) 40 km   b) 24 km

Exercise 7b   p. 63

1. a) 72 p   b) 30 p   **3.** 19 tonnes   **5.** a) 10 miles   b) 30 miles
2. a) £ 10   b) £ 13   **4.** a) £ 11.50   b) £ 23   **6.** a) 16 tonnes   b) 21 tonnes

Exercise 7c   p. 64

1. a) £ 2.80   b) £ 5.80   **3.** a) £ 50   b) £ 30   **5.** a) Nothing   b) 20 p
2. a) 110 p or £ 1.10   b) £ 1.80   **4.** a) 6   b) 9   c) 38 p   d) 32 p

Exercise 7d   p. 65

1. 12   **6.** 0   **11.** 10
2. 18   **7.** 15   **12.** 34
3. 5   **8.** 2   **13.** a) 32   b) 52
4. 13   **9.** 33   c) 8
5. 15   **10.** 6   **14.** a) 2   b) 3

Exercise 7e   p. 66

1. 5	**6.** 66	**11.** 5
2. 10	**7.** 12	**12.** 50
3. 12	**8.** 5	**13.** 12
4. 6	**9.** 9	**14.** 33
5. 11	**10.** 14	

Exercise 7f   p. 67

In the worked example, the value of b is intended to be found intuitively and not by a formal solution of the equation.

1. 4	**4.** 12	**7.** a) 40°
2. 4	**5.** 6	b) 60°
3. 2	**6.** 90°	**8.** 80°

Exercise 7g   p. 67

1. 41	**3.** £6.40	**5.** 3
2. 7	**4.** £10	**6.** 1.8

CHAPTER 8   Lines and Angles

Basic angle work is covered in this chapter; it is covered fairly briefly because we assume that all that is needed at this stage is reinforcement. If a more thorough approach is needed, there is plenty of material in ST(P) Mathematics Book 1.

Almost all practical work, from the designing and making of simple models or real objects (toys, models, furniture, etc.) to the maintenance of property from bicycles to houses, needs an understanding of angles and how they are measured.

Exercise 8a   p. 69

1. 4	**4.** a) 10	b) 15	**5.** a) 15 s	b) 45 s
2. $\frac{3}{4}$	c) 60	d) 900	c) 30	
3. 1 right angle (or 90°)				

Exercise 8b   p. 70

1. a) $\frac{1}{2}$	c) $\frac{1}{6}$	**4.** a) $\frac{1}{4}$	c) $\frac{1}{8}$	**6.** a) 90°	c) 15°
b) $\frac{1}{4}$	d) $\frac{1}{12}$	b) $\frac{1}{3}$	d) $\frac{1}{6}$	b) 30°	d) 180°
2. a) 180°	c) 60°	**5.** a) $\frac{1}{4}$	c) $\frac{1}{24}$	**7.** 75°	
b) 90°	d) 30°	b) $\frac{1}{12}$	d) $\frac{1}{2}$	**8.** a) 360°	c) 120°
3. 270°				b) 180°	d) 360°

Exercise 8c   p. 71

1. Acute
2. Obtuse
3. Obtuse
4. Acute
5. Obtuse
6. Acute
7. Acute
8. Obtuse
9. Obtuse
10. a)

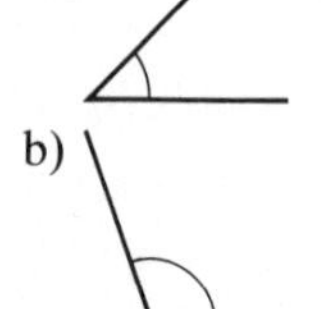

b)

c)

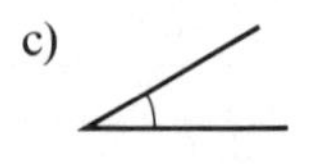

d)

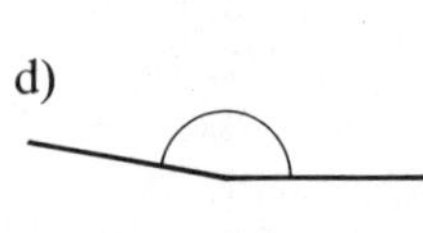

e)

f)

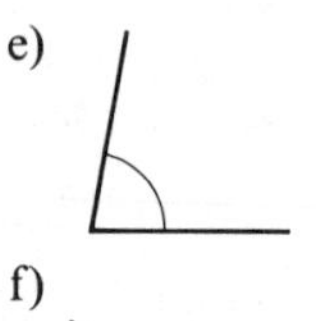

Exercise 8d   p. 73 —————————————————————————————

Skill at estimating the size of an angle is very useful. Plenty of practice in both measuring and estimating angles is needed: a duplicated sheet containing just angles can be made easily and is ideal for this purpose.

1. b) 24°

2. b) 144°

3. b) $p = 43°, q = 137°$

c) should be about 180°

4. b) $d = 90°, e = 114°, f = 58°,$

$g = 34°$

Exercise 8e   p. 74 —————————————————————————————

1. 110°

2. 165°

3. 70°

4. 65°

5. 50°

6. 105°

7. 45°

8. 75°

9. 107°

10. 35°

11. 20°

12. 108°

Exercise 8f   p. 77 —————————————————————————————

1. a) DC

b) AD (or BC)

c) AD

2. a) AG is parallel to DE and CF.

b) AB is parallel to DC and EF.

c) DE is perpendicular to EF and DC and AB.

3. a) True

b) False

c) True

d) False

Exercise 8g   p. 78 —————————————————————————————

1. 25°

2. a) 1080°   b) 15

c) 15 minutes

3. 240°

4. a) 44°   b) 136°

c) 180°

5. a) BD   b) DE

c) BC   d) AB and DE

e) 90°

Exercise 8h   p. 79 —————————————————————————————

1. 122°

2. $\frac{2}{3}$

3. b) about 142°   c) 142°

4. a) FD   b) BE

c) no   d) about 45°

e) 53°

5. a) 360°   b) 180°

c) 90°   d) 45 seconds

e) 10 minutes

CHAPTER 9   Triangles and Quadrilaterals

This continues the reinforcement of basic geometry, concentrating on shape and the measurement of lines and areas. It also revises the sum of the angles in triangles and quadrilaterals. Practical work on this chapter could include projects on furnishing or decorating that need areas and perimeters to be calculated, and model making or designing in which angles need to be worked out as well as lengths and areas.

Exercise 9a   p. 82 —————————————————————————————

1. a) AB = 8.2 cm, BC = 5.7 cm, AC = 10 cm

b) 24 cm

c) Angle A = 35°, angle B = 90°, angle C = 55°

d) 180°

2. a) PQ = 8.6 cm, QR = 5.6 cm, PR = 13.5 cm

b) 27.7 cm

c) Angle P = 14°, angle Q = 144°, angle R = 22°

d) 180°

3. a) AB = 6.5 cm, BC = 6.5 cm, AC = 11.2 cm, perimeter = 24.2 cm

b) 90°, yes

Exercise 9b   p. 84

(Diagrams are half-size.)

1. a) AB = 10.1 cm, BC = 7.1 cm,
   CD = 10.8 cm, AD = 3.7 cm
   b) 31.7 cm
   c) Angle A = 74°, angle
      B = 101°, angle C = 61°,
      angle D = 125°
   d) 360°

2. a)
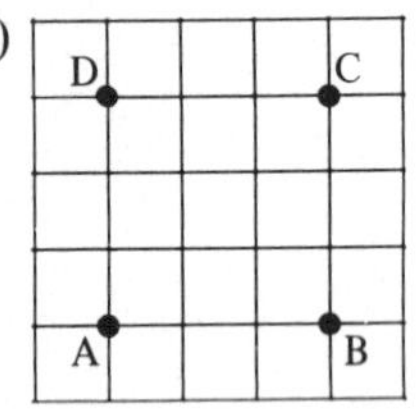

   b) 12 cm   d) 4.2 cm

3. a) 20 cm
   b)
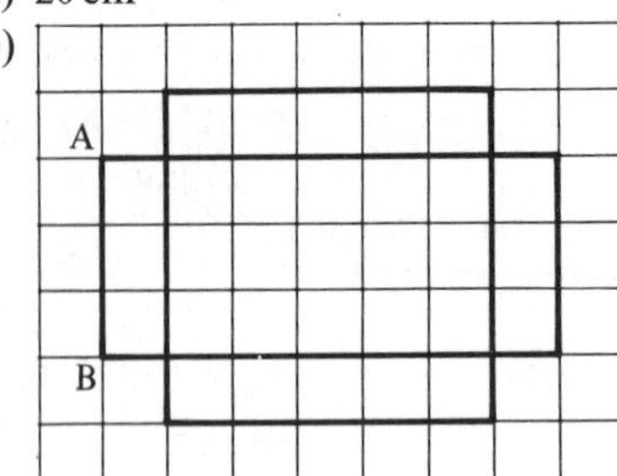

4. a) 36 cm
   b) $\widehat{A} = \widehat{B} = 90°$, $\widehat{C} = 53°$
      $\widehat{D} = 127°$; 360°
   c) 14.4 cm   d) 9 cm

5. a) 24 cm   b) 8 cm, 4 cm and 4 cm

Exercise 9c   p. 87

1. 12 cm²
2. 12 cm²

3. 17 cm²
4. 8 cm²

Exercise 9d   p. 88

After this exercise has been worked, the following investigation can be done. The number of sticks to be used can be varied: it does not have to be a square number.

Material needed: several sheets of 1 cm dotted grid paper
                   a supply of 1 cm long sticks (thin balsa wood is ideal, but any material that can be easily cut will do). These are not essential, but will help understanding.

Problem 1   a) Draw five different shapes that can be enclosed by 36 sticks.
            b) Find the area of each of your shapes.
            c) Find the maximum area that you can enclose with your 36 sticks.
            d) Find the smallest area that can be enclosed by your 36 sticks.

Problem 2   a) Draw five different shapes that have an area of 25 cm².
            b) Find the number of sticks needed to enclose each of your shapes.
            c) Find the largest number of sticks that can be used to enclose an area of 25 cm².
            d) Find the smallest number of sticks that can be used to enclose an area of 25 cm².

1. 10.32 cm²
2. 1971 cm²
3. 546 cm²

4. 361 cm²
5. 42 000 cm²
6. 3672 cm²

7. a) 8 cm²   d) 6 cm²
    b) 5 cm²   e) 6 cm²
    c) 7 cm²   f) 7 cm²

Exercise 9e   p. 91

1. 4 cm²
2. 6 cm²
3. 3 cm²

4. 4 cm²
5. 5 cm²
6. 7.5 cm²

7. 2.5 cm²
8. 6 cm²

Exercise 9f   p. 92

1. 4 m²
2. 6 m²
3. 15 sq yd
4. 10.5 sq yd

5. 1.5 m²
6. 20.25 sq yd
7. 315 m²
8. 58 m²

9. a) A: 1500 m², B: 700 m², C: 450 m²,
      D: 50 m², E: 50 m², F: 400 m²
    b) 3150 m²

Exercise 9g   p. 94

1. a) 4 cm
   b) 16 cm
2. a) 5 cm
   b) 25 cm²

3. a) 50 m
   b) 200 m
4. A: 6 cm², B: 3 cm², C: 4.5 cm²,
      D: 4 cm², E: 8 cm²

5. a) 36 cm²   c) 18 cm²
   b) 4.5 cm²   d) 18 cm²
6. a) 3 cm²   b) B, C, D, F, G
7. a) 16 cm²   b) 8 cm

Exercise 9h   p. 97

1. 45°
2. 22 °
3. 20 °

4. 50°
5. 33°
6. 119°

7. 75°
8. 31°

Exercise 9i   p. 98

1. 90°
2. 45°
3. 90°

4. 130°
5. 50°
6. 30°

7. 105°
8. 115°

Exercise 9j   p. 99

1. 40°
2. 90°
3. 45°

4. 60°
5. 103°
6. 90°

7. 47°
8. 50°

Exercise 9k   p. 100

1. $x = 132°, y = 70°$
2. $p = 60°, q = 60°$

3. $s = 40°, t = 140°$
4. $f = 50°, g = 50°$

5. $m = 65°, n = 65°$
6. $s = 140°, t = 45°$

Exercise 9l   p. 101

1. 75°
2. a) i) 60 cm   ii) 96 cm²   iii) 53°
   b) 74°
3. a) i) 18.5 cm   ii) 12.4 cm²
   b) 26°
   c) 64°

4. 30°
5. a) i) 8.6 cm²   ii) 32.1 cm²   iii) 31° ’
   c) 59°

CHAPTER 10   Symmetry

The work in this chapter includes the use of symmetry for deducing properties of figures. The provision of some everyday objects that have the different types of symmetry discussed here can be helpful. This idea is continued in Chapter 20, where special triangles and quadrilaterals are investigated. These two chapters can be studied together.

Exercise 10a   p. 104

1.
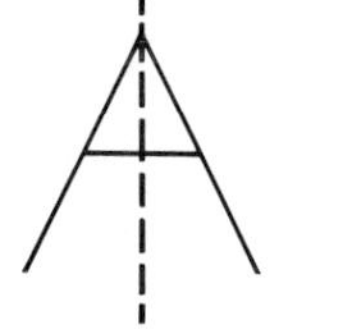

3.
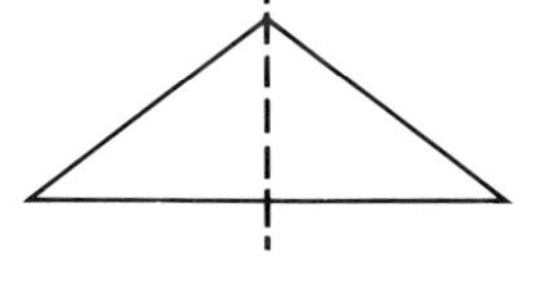

5.
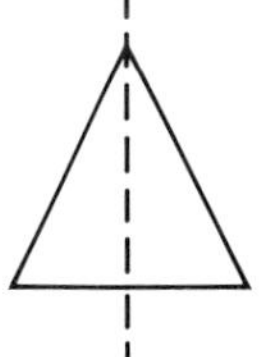

2.
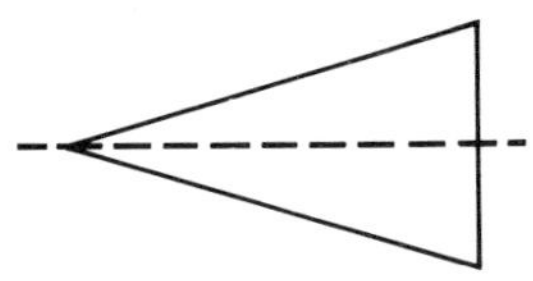

4.
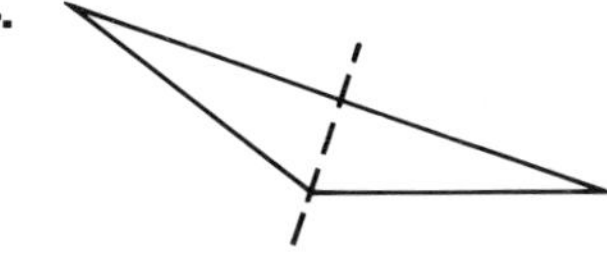

6. 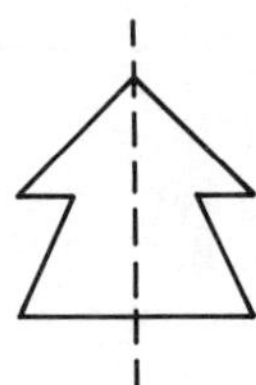

7. 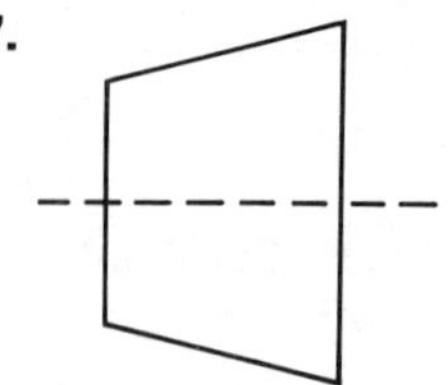

8. 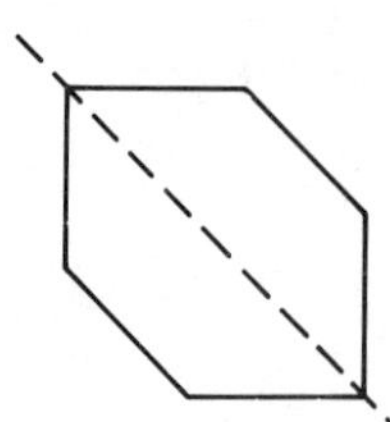

Exercise 10b   p. 106

1. 2 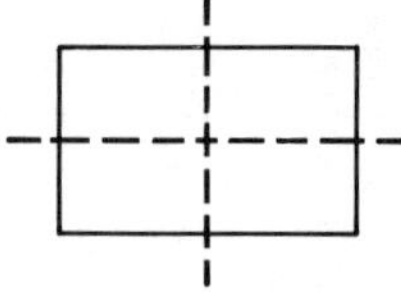

2. 2 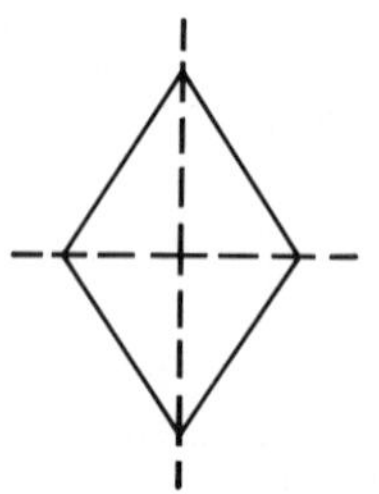

3. 4 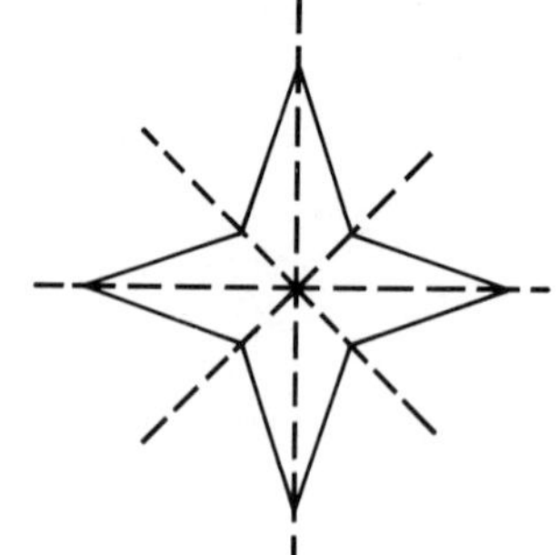

4. 2 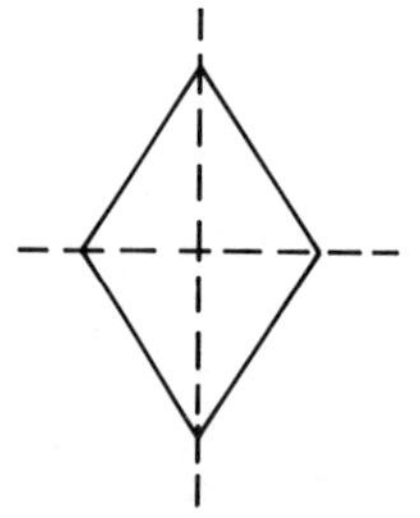

5. 2 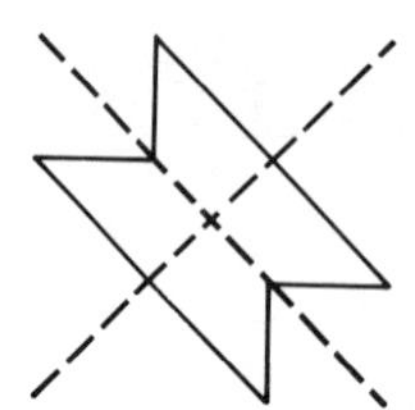

6. 3 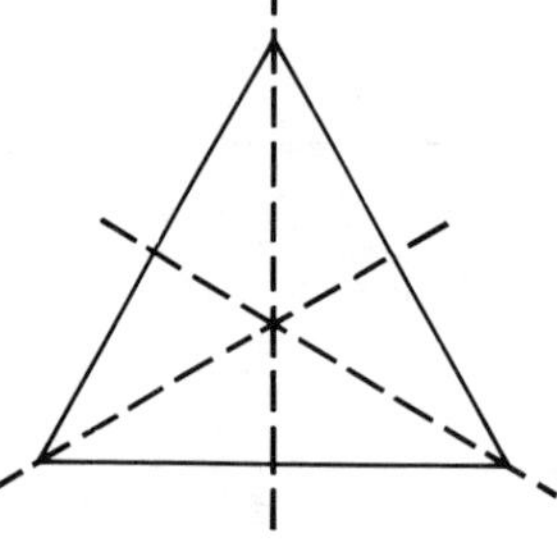

7. 2

8. 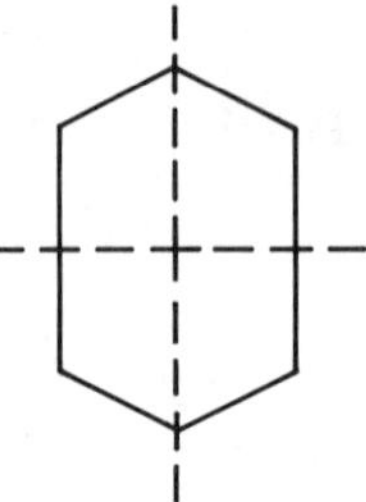

9. 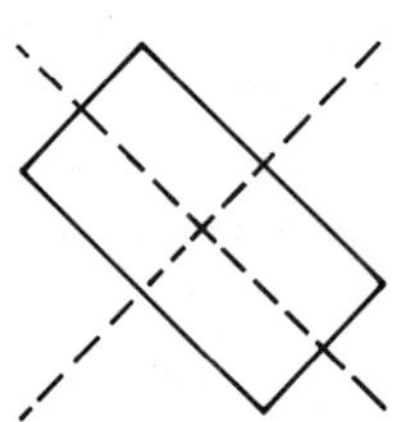

10. 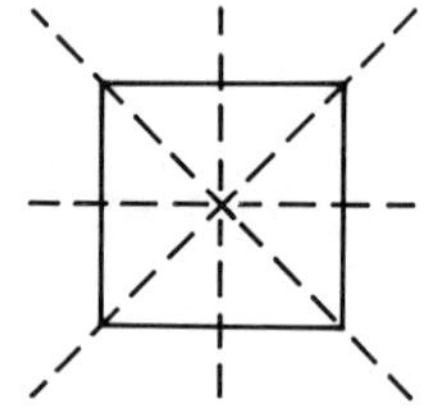

Exercise 10c   p. 108

1. 2
2. 2
3. 3
4. 4
5. 3
6. 2
7. 2
8. 2
9. 3
10. 4

Exercise 10d   p. 109

1. a) 4   b) 4
2. a) None   b) None
3. a) 1   b) None
4. a) 2   b) 2
5. a) None   b) 2
6. a) None   b) 3
7. a) 2   b) 2
8. a) 4   b) 4
9. a) 6   b) 6

Exercise 10e   p. 110

1. a) AC = BC   b) $\widehat{A} = \widehat{B}$
2. a) PQ = PR   b) $\widehat{Q} = \widehat{R}$
3. a) True   c) True
    b) False   d) False

4. a) True   e) False
    b) True   f) True
    c) True   g) True
    d) False   h) False

5. a) False   e) True
    b) True   f) False
    c) True   g) False
    d) False   h) True

Exercise 10f   p. 111

1. a) True   d) True
    b) False   e) True
    c) True   f) False
2. a) $\widehat{A} = \widehat{C}$ c) AB = CD
    b) $\widehat{B} = \widehat{D}$ d) BC = DA

3. a) True   c) False
    b) True
4. a) AB = BC = CA
    b) $\widehat{A} + \widehat{B} + \widehat{C}$ = 180 degrees
    c) $\widehat{A} = \widehat{B} = \widehat{C}$ = 60 degrees

5. a) True   c) True
    b) False   d) True

Exercise 10g   p. 112

1. a) 1
    b)

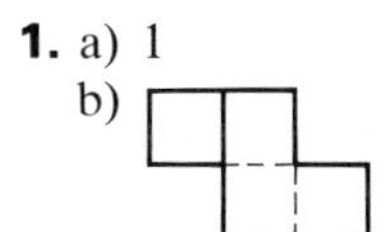

2. a) 6   b) 360°
    c) 60°

3. a) 100°   b) 360°
    c) 120°

4. a) 2   b) 360°
    c) 55°   d) 125° each
5. 23 cm
6.

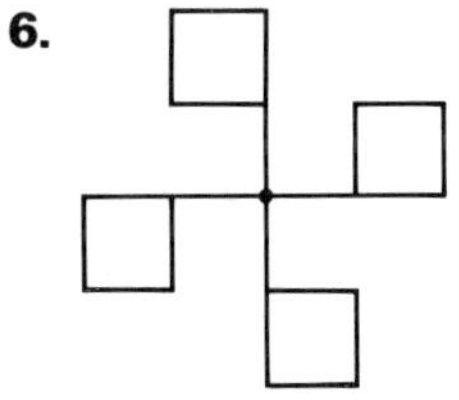

CHAPTER 11   Solids

We recommend that each solid should be made before the relevant exercise is worked. If possible the models should be kept, as they will be useful in Exercise 11e. A plentiful supply of small cubes will help with some questions, particularly in the section on volumes. (Sugar cubes will do and Lego bricks are useful as they can be made up into the shapes required.) Models made of wire showing just the edges, are also quite useful. An extended piece of work following on from this chapter could be based on designing a 'fit together puzzle' based on a 4 cm cube.

Exercise 11a   p. 115

2. 6
3. 12

4. 8
5. Square

Exercise 11b   p. 116

2. 6
3. 12

4. 8
5. Rectangular

6. a) 2
    b) 4

Exercise 11c   p. 118

1. Cuboid
3. 5
4. 9
5. 6

6. 3
7. Triangular; 2
8. a) 8   b) 18   c) 12
    d) 6   e) Hexagonal

9. a) 7
    b) 15
    c) 10

Exercise 11d   p. 120

1. 5
2. 8
3. 5
4. 4

5. 4
6. 6
7. 4
8. Equilateral triangle

9. a) 4
    b) Square or rectangular
    c) 5
    d) 8

Exercise 11e   p. 121

1. a) Cuboid   b) 6
  c) 12
2. a) Hexagonal prism
  b) 8   c) 18
3. a) Triangular prism
  b) 5   c) 9

4. a) Cuboid
  b) 6   c) 12
5. 4
6. a) Cuboid   b) 5
7. a) 8   b) Square
  c) 8   d) 12

8. a) 6   b) 3
9. a) 6   b) 9
  c) 12
10. a) 8   b) 3

Exercise 11f   p. 124

1. a) 27   b) 6
  c) 1

2. a) 2   b) 2
3. a) 17   b) 4

Exercise 11g   p. 125

1.

3. 1
4. 1

5. a) 4   b) 1
6. a) B   e) C
  b) A   f) D
  c) C   g) D
  d) B

7. 1st and 3rd
8. 1st and 2nd

Exercise 11h   p. 128

1. $8\,cm^3$
2. $12\,cm^3$

3. $10\,cm^3$
4. $27\,cm^3$

5. $24\,cm^3$
6. $20\,cm^3$

Exercise 11i   p. 128

1. $4\,cm^3$
2. $6\,cm^3$
3. $12\,cm^3$

4. $8\,cm^3$
5. 80
6. 12

7. a) 20
  b) $8\,cm^3$
  c) $160\,cm^3$

Exercise 11j   p. 130

1. $40\,cm^3$
2. $12\,cm^3$
3. $120\,cm^3$

4. $1440\,cm^3$
5. $3750\,cm^3$
6. $18\,cm^3$

7. a) 162 cm   d) $288\,cm^2$
  b) $8640\,cm^3$ e) 6
  c) $720\,cm^2$
8. a) $1920\,cm^3$ b) $800\,cm^2$

CHAPTER 12   Units

Many of the units dealt with in this chapter appear as meaningless names to many pupils. This can be overcome to some extent if concrete objects can be produced when discussing particular units. Estimating larger quantities from actual objects also helps to give some idea of size. For example they could be asked to estimate the length of the room they are in, with a metre rule to look at. They could also be asked to draw a line 8 cm long, say, without looking at a ruler, and then measure it to see how accurate they have been.

Exercise 12a   p. 132

1. 3000 m
2. 1200 mm
3. 4.5 km
4. 50 cm

5. 12 cm
6. 7.5 m
7. 2600 m
8. 3.4 km

9. 30 cm
10. 500 m by 250 m

Exercise 12b   p. 134

1. $500\,mm^2$
2. $5000\,cm^2$
3. $8\,m^2$
4. $8\,cm^2$

5. $30\,000\,m^2$
6. $2400\,mm^2$
7. $13\,000\,cm^2$
8. 1.5 hectares

9. £3.50
10. $0.2\,m^2$, £1.60

Exercise 12c   **p. 135**

1. 3000 ml
2. 40 000 cm³
3. 5000 mm³
4. $\frac{1}{4}$
5. 0.5 cm³
6. 27 000 mm³
7. 530 000 cm³.
8. 750 ml
9. 1500 ml
10. 0.75 litres
11. $22\frac{1}{2}$ litres

Exercise 12d   **p. 136**

1. 2000 g
2. 0.5 kg
3. 1200 kg
4. 4.2 t
5. 1.26 t
6. 400 g
7. 50 000 kg
8. 1.6 kg
9. 6 t
10. 40

Exercise 12e   **p. 138**

1. 9 ft
2. 98 lb
3. 64 pints
4. 12 ft
5. 9 yd
6. 4.5 lb
7. 5280 yd
8. 3 cu yd
9. 5.25 gallons
10. 484 sq yd
11. 60 inches
12. 1296 sq inches

Exercise 12f   **p. 139**

This exercise illustrates some everyday situations in which conversion between metric and imperial units may arise. The pupils can be asked to provide some examples themselves. This exercise can also be used to discuss how precise the answers to this type of problem should be. It is worth pointing out the ridiculousness of using 100 g = 3.5 oz to convert, say, 2 oz into grams and giving the answer from a calculator as 57.142857 g.

Questions based on this exercise can be used for aural work. Encourage conversion without a calculator: shopping expeditions frequently require this facility.

1. 64 km
2. $62\frac{1}{2}$ miles
3. 20 cm
4. 78 in
5. 17.5 pints
6. 10.5 oz
7. 150 acres
8. 24 in
9. 600 g
10. 230 g
11. 18 litres
12. 225 ml
13. 55 lb
14. 36 litres
15. 9 gallons
16. 1.8 g
17. 54 litres
18. 6
19. 6
20. a) 17.9 sq yd (to nearest $\frac{1}{10}$ th yd)
    b) £ 75
    c) £ 76.08 if sold to the nearest $\frac{1}{10}$ of a yard

Exercise 12g   **p. 140**

1. 6 km/h
2. 360 m.p.h.
3. 2 m/s
4. 6 m.p.h.
5. 84 km/h
6. 5 m.p.h.
7. 1250 m.p.h.
8. 4 m/s
9. 9 m.p.h.
10. 50 m.p.h.

Exercise 12h   **p. 141**

1. 8 km/h
2. 7 m.p.h.
3. 43 m.p.h.
4. 400 km/h
5. $8\frac{1}{3}$ m/s

Exercise 12i   **p. 142**

1. 112 km/h
2. 31.25 m.p.h.
3. 160 km/h
4. 50 m.p.h.
5. 281.25 m.p.h.
6. $62\frac{1}{2}$

Exercise 12j   **p. 143**

1. 80 m.p.g.
2. 36 m.p.g.
3. 4 km/litre
4. 5 km/litre
5. 4 gallons
6. a) 60 m.p.g.   b) 2.5 gallons
7. a) 800 km   b) 60 litres
8. 216 miles
9. a) 16 litres/min
   b) 192 litres   c) 10 min
10. a) 25 ml per shampoo
    b) 125 ml   c) 3

11. $60\,\text{m}^2$

12. a) 4 tablets per day
b) 15

13. a) 48   b) 3

14. a) i) 5 gal/min   ii) 300 gal/h
b) 4 h

15. a) $16\,\text{m}^2$/litre
b) 7.75 litres   c) $240\,\text{m}^2$

CHAPTER 13   Nets

The solids made for chapter 11 will be useful for this chapter. A plentiful supply of 1 cm squared paper, scissors and adhesive tape are needed for much of the work. It will also be helpful if some paper models of cubes etc. can be made up in advance, ready to open out for questions such as no. 3 of Exercise 13a. The edges can be held together with ordinary small sticky labels which will peel off a reasonably shiny paper: these solids need to be put together and opened out several times.

Exercise 13a   p. 147 ─────────────────────────────

1. These are the nets of a cube:

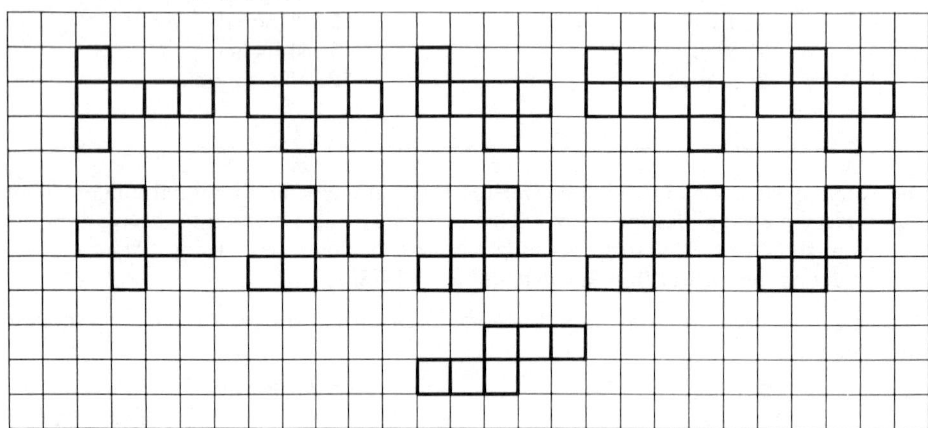

These are some arrangements that are not nets:

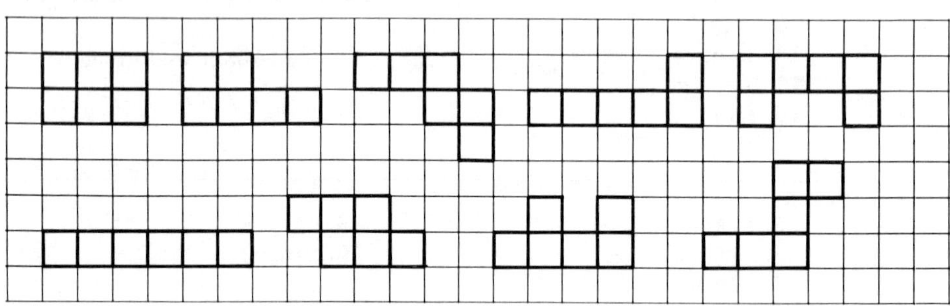

2. a) IH   b) D and B

3.

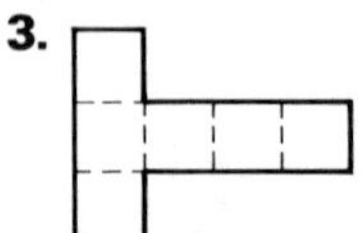

Exercise 13b   p. 148 ─────────────────────────────

1. a) i) 2   ii) 2
iii) Each is 4 cm by 3 cm

b) One possibility is

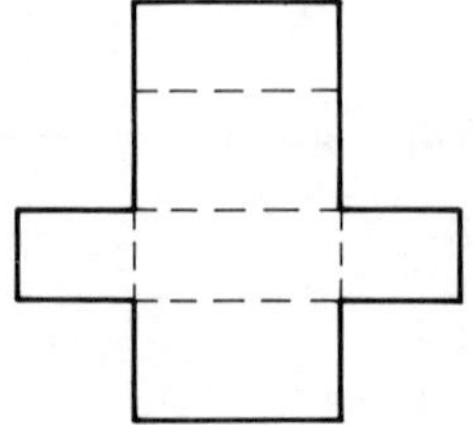

2. a) 6

b) 2 each of

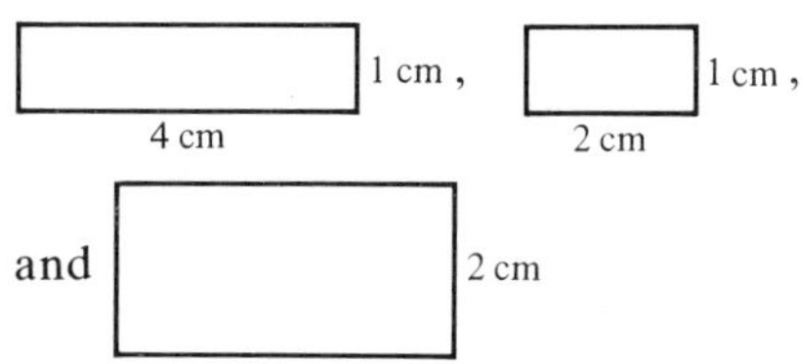

and

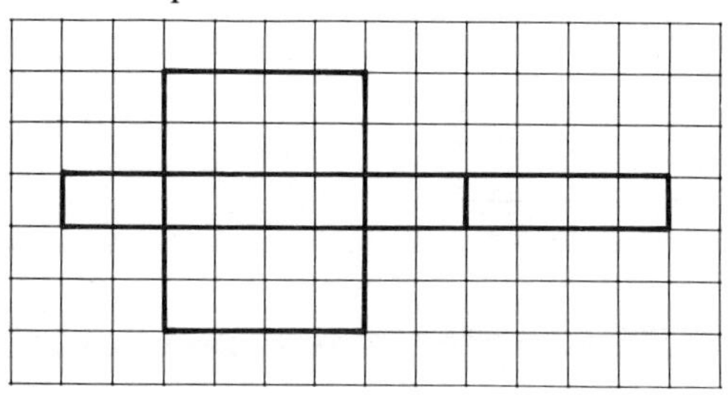

c) One example is

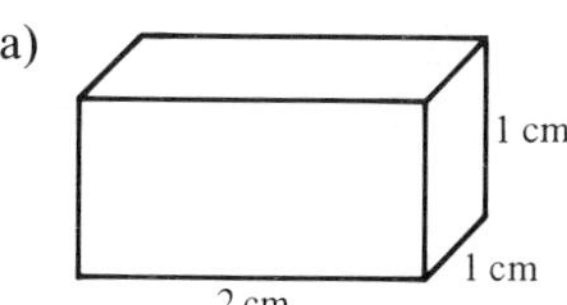

3. a)

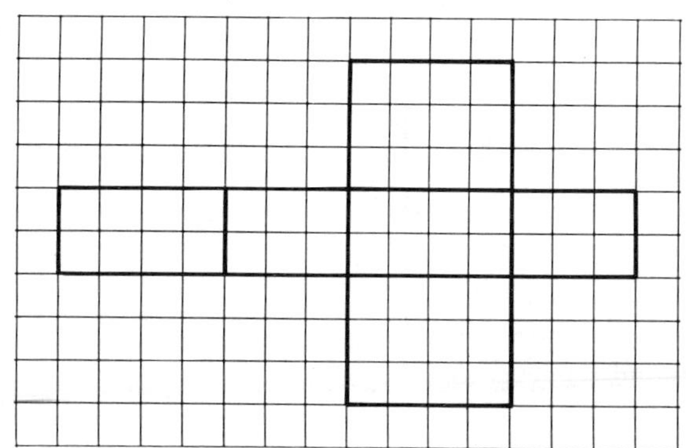

b) IJ   c) K and G

4.

Exercise 13c   p. 149

1. a) i) 3, no
   ii) 2, same shape and
      right-angled
b) One net is

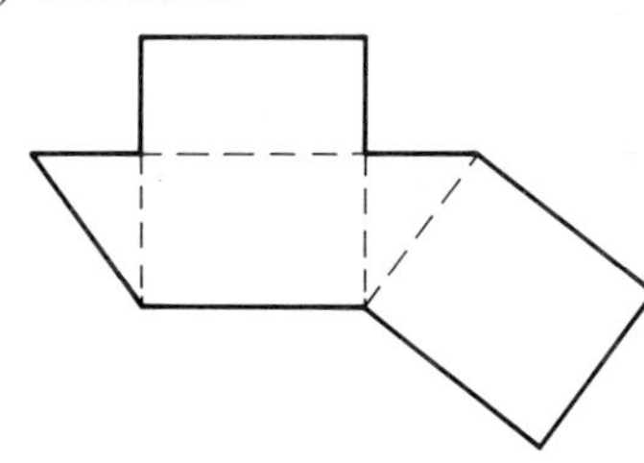

2.

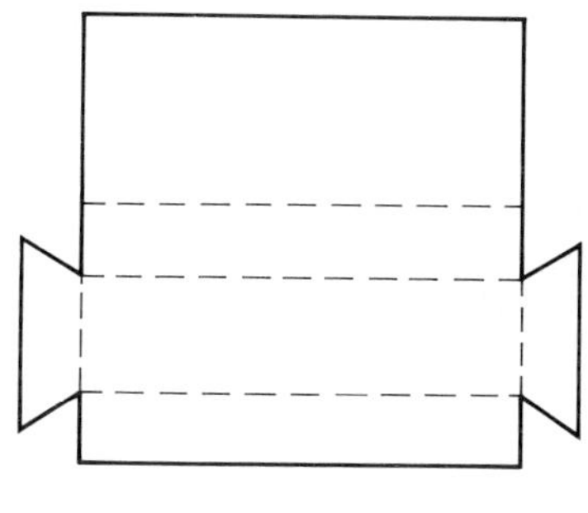

3. a) Triangular
   b) DE   c) A, G

Exercise 13d   p. 150

1. a) Cube
   b) Cuboid

c) Square pyramid
d) Triangular prism

e) Cone
f) Triangular pyramid (tetrahedron).

2. a)

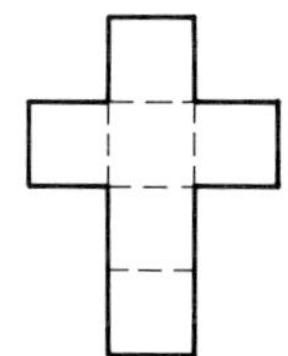

c)

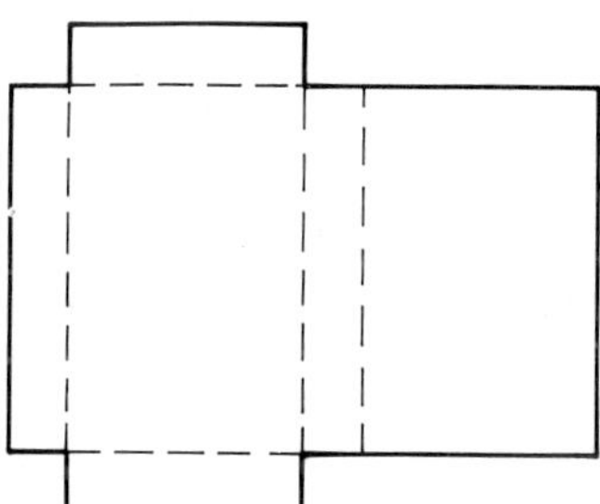

b)

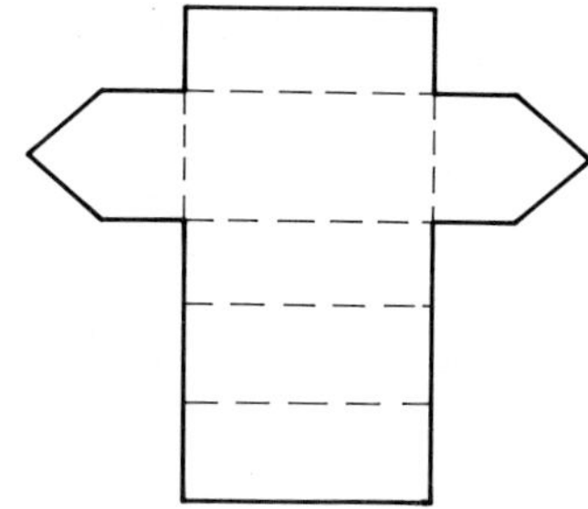

d)

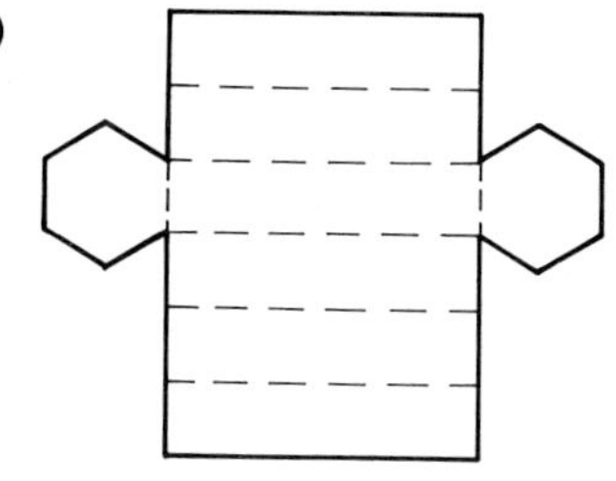

3. a)

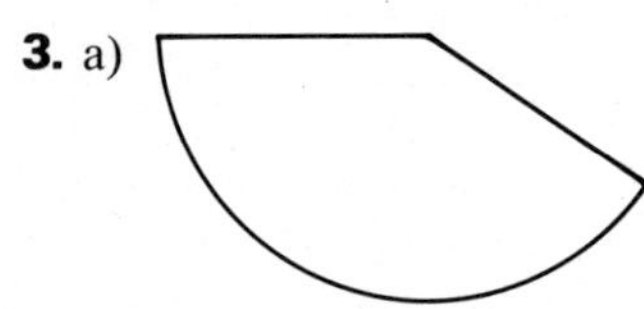

b)

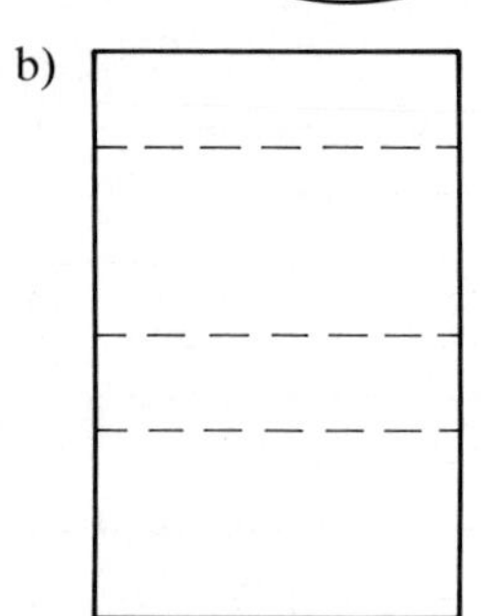

c)

d)

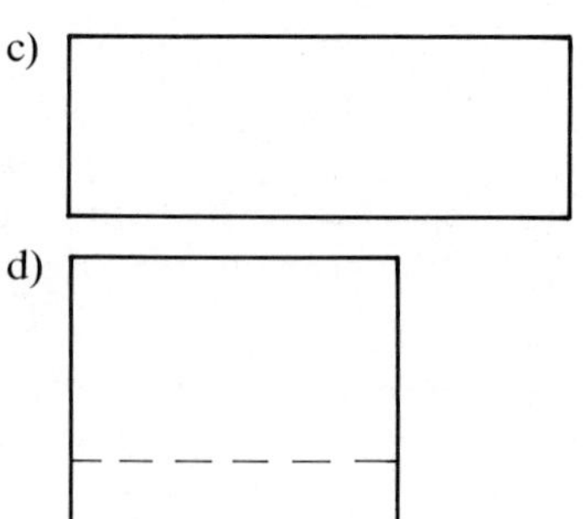

4. A and B

CHAPTER 14   Positive and Negative Numbers

This chapter uses positive and negative numbers in practical situations.

Exercise 14a   p. 152

1. $2\,°C$
2. $-5\,°C$
3. a) $20\,°C$ b) $0\,°C$
   c) $-10\,°C$ d) $-4\,°C$
4. $-8\,°C$
5. $30\,°$
6. $0\,°$
7. $0\,°$

8. $16\,°$
9. $-12\,°$
10. $-7\,°$
11. $10\,°$
12. $8\,°$
13. $5\,°$
14. $25\,°$
15. $-15\,°$

16. $-15\,°$
17. $-1\,°$
18. $-8\,°$
19. $-11\,°$
20. $0\,°$
21. $-9\,°$
22. $-20\,°$

Exercise 14b   p. 153

1. $8\,°C$
2. $4\,°C$
3. $4\,°C$
4. $8\,°C$
5. $4\,°C$

6. $4\,°C$
7. $3\,°C$
8. $4\,°C$
9. $12\,°C$
10. $12\,°C$

11. $11\,°C$
12. $6\,°C$
13. $-10\,°C$
14. a) $10\,°C$ b) $11\,°C$
   c) $-2\,°C$

Exercise 14c   p. 154

Some newspapers publish world temperature charts each day and these can be used for further examples.

1. a) No   b) Moscow
   c) London
   d) $-9\,°, -4\,°, 1\,°, 3\,°, 4\,°$
2. a) January   b) October
   c) $-4\,°, -3\,°, -2\tfrac{1}{2}\,°, 0\,°, \tfrac{1}{2}\,°, 6\,°$
   d) $10\,°C$ e) $8\tfrac{1}{2}\,°C$

3. a) All except Berlin
   b) Berlin, Belgrade, Bordeaux, Biarritz, Barcelona
   c) Warmer   d) $10\,°C$
   e) $12\,°C$

Exercise 14d   p. 155

1. $-2, -5, -10$
2. $-1, -2, -5$
3. 3
4. $5, -1, -8$
5. a) $-5, -3, -1, 0, 4$
   b) $-7, -5, -2, -1, 2, 7$

6. a) $-6, -4, 3, 7$
   b)

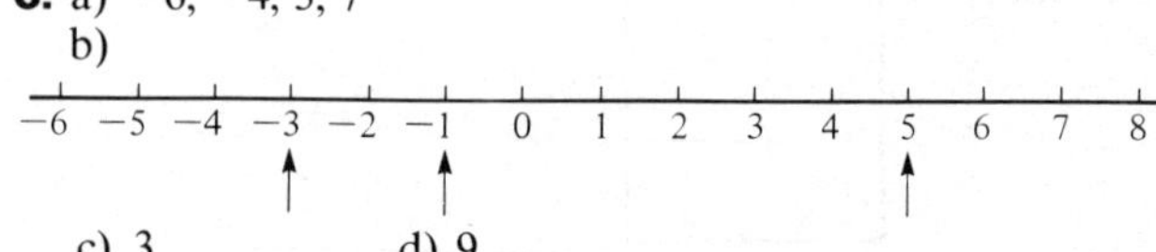

   c) 3   d) 9

Exercise 14e   p. 156

1. a) 4th   d) −3
    b) 3rd   e) 4th
    c) +2
2. a) He has a debt to pay.
    b) Meg   d) £6
    c) Jim   e) −60 p

3. a) No   d) £11 D
    b) −£45, yes   e) £17 C
    c) £62 C
4. a) Monday, Friday, Saturday
    b) Amount paid out
    c) Profit

5. a) Forward   d) 1 pace
    b) Backwards   e) 5 paces
    c) 8 paces   f) 3 paces
6. a) 210 m   c) 18 m
    b) 92 m

CHAPTER 15   Circles

This topic can be introduced by showing a circle of about one foot diameter. Suggest that it is a table top that needs veneering round the edge and ask what length of veneer is needed. (Point out that veneer is not cheap and buying too much or too little is going to be an expensive mistake!)

Exercise 15a   p. 160

3. c) 10 cm
    d) Diameter = 2 × radius

4. 12 cm
5. 7 cm

Exercise 15b   p. 162

1. 12 m
2. 45 cm
3. 24 cm

4. 150 cm
5. 24 inches
6. 4.5 m

7. a) 150 cm
    b) 150 cm
    c) 30 m

Exercise 15c   p. 163

Point out that, as 3.14 is an approximate value for π, answers with four or five decimal places cannot be accurate so it is silly to copy all the figures given by a calculator. The answers here are given correct to one decimal place.

1. 31.4 cm
2. 29.5 cm
3. 138.2 cm
4. 100.5 cm

5. 15.7 m
6. 37.7 ft
7. 25.1 cm
8. 78.5 m

9. 24.5 m
10. 10.0 cm
11. Obviously wrong. Rough
    answer is 33 cm.

Exercise 15d   p. 164

1. 87.9 cm
2. 37.7 cm
3. a) 28.3 cm
    b) 29.3 cm
4. 31.4 ft

5. 314 m
6. a) 1.9 m   b) 1.9 m
    c) 37.7 m   d) 38 m
7. a) 25.1 in   b) 251.2 in
    c) 250 in   d) 20 times

8. a) 31.4 cm, 62.8 cm
    b) 94.2 cm
    c) 109.2 cm
    d) 110 cm

CHAPTER 16   Probability

It is easy for pupils to get the impression from text books such as this one that probability is concerned only with winning and losing games. It is worth discussing the relevance of probability in everyday life, from simple decisions such as 'Shall I take an umbrella out with me?' to the applications of probability in areas such as insurance, planning for future numbers in education, sampling items from production lines to judge the quality of all the output, and so on.

Projects that could follow from this work include investigations on fruit machines (e.g. what is the chance of winning the jackpot?), roulette and setting up their own 'book' on a game. Investigations that compare experimental probability with theoretical probability can be interesting. For example, have a bag with five different coloured discs (or Lego bricks, etc.) and pick one out. Repeat 50 times and keep a record of the colour taken out. From this record what is the probability of choosing a red disc? How does this compare with the theoretical probability? If there is a big difference between the two results, try to explain it.

It can be interesting to compare probabilities of real-life events, e.g. the probability of being murdered is much greater than the probability of winning on the pools!

Exercise 16a   p. 167

1. a) $\frac{1}{6}$ b) $\frac{1}{6}$ **3.** $\frac{1}{4}$

    c) $\frac{1}{2}$ **4.** $\frac{1}{2}$

2. a) $\frac{1}{6}$ b) $\frac{1}{6}$ **5.** $\frac{1}{5}$

    c) $\frac{1}{3}$ **6.** $\frac{2}{5}$

Exercise 16b   p. 168

1. 0   **3.** a) $\frac{1}{2}$ b) $\frac{1}{4}$

2. 1   **4.** $\frac{1}{40}$

Exercise 16c   p. 168

Some of these questions are quite difficult.

1. a) $\frac{1}{4}$ b) $\frac{1}{5}$ **6.** 20   **9.** a)

    c) Matthew. $\frac{1}{4} > \frac{1}{5}$ **7.** $\frac{19}{100}$

×	0	1	2	3
1	0	1	2	3
2	0	2	4	6
3	0	3	6	9

2. a) $\frac{3}{5}$ b) $\frac{1}{2}$ **8.** a)

3. a) $\frac{1}{15}$ b) $\frac{1}{10}$

+	4	5	6
1	5	6	7
2	6	7	8
3	7	8	9

    c) $\frac{5}{30}$ i.e. $\frac{1}{6}$

b) $\frac{2}{12}$ i.e. $\frac{1}{6}$

c) $\frac{3}{12}$ i.e. $\frac{1}{4}$

4. a) $\frac{1}{4}$ b) 6

d) $\frac{4}{12}$ i.e. $\frac{1}{3}$

5. a) $\frac{4}{12}$ i.e. $\frac{1}{3}$ b) $\frac{2}{12}$ i.e. $\frac{1}{6}$ b) $\frac{1}{9}$ c) $\frac{3}{9}$ i.e. $\frac{1}{3}$

    c) $\frac{6}{12}$ i.e. $\frac{1}{2}$ d) Total 1   d) $\frac{5}{9}$

CHAPTER 17   Working with Percentages

The basic percentage processes are revised in this chapter. There are a few problems on the application of percentages but the use of percentages in everyday contexts follows throughout the rest of the book.

This chapter can be used as necessary for reminders in the chapters that follow.

Exercise 17a   p. 171

1. 0.2   **17.** 3.1   **30.** $\frac{1}{10}$

2. 0.8   **18.** 1.5

3. 0.5   **19.** 2.75   **31.** $\frac{3}{10}$

4. 0.7   **20.** 0.84   **32.** $\frac{4}{5}$

5. 0.9   **21.** 1.37

6. 0.6   **22.** 0.58

7. 0.3   **23.** 1.56

8. 0.1   **24.** 0.32

9. 0.45   **25.** $\frac{1}{4}$

10. 0.75   **26.** $\frac{3}{20}$

11. 0.39

12. 0.58   **27.** $\frac{1}{2}$

13. 0.25

14. 0.98   **28.** $\frac{1}{5}$

15. 0.35

16. 0.18   **29.** $\frac{3}{10}$

	Percentage	Fraction	Decimal
33.	20%	$\frac{1}{5}$	0.2
34.	25%	$\frac{1}{4}$	0.25
35.	50%	$\frac{1}{2}$	0.5
36.	75%	$\frac{3}{4}$	0.75
37.	90%	$\frac{9}{10}$	0.9
38.	130%	$\frac{13}{10}$ or $1\frac{3}{10}$	1.3
39.	175%	$\frac{7}{4}$ or $1\frac{3}{4}$	1.75

Exercise 17b   p. 173

1. 25%
2. 30%
3. 70%
4. 95%
5. 82%
6. 46%
7. 29%
8. 65%

9. 140%
10. 125%
11. 280%
12. 375%
13. 25%
14. 40%
15. 24%
16. 10%

	Percentage	Decimal	Fraction
17.	50%	0.5	$\frac{1}{2}$
18.	75%	0.75	$\frac{3}{4}$
19.	25%	0.25	$\frac{1}{4}$
20.	80%	0.8	$\frac{4}{5}$
21.	85%	0.85	$\frac{17}{20}$
22.	37.5%	0.375	$\frac{3}{8}$
23.	35%	0.35	$\frac{7}{20}$

Exercise 17c   p. 173

1. 80%
2. 60%

3. 60%
4. $\frac{3}{4}$

5. $\frac{3}{20}$
6. a) $\frac{1}{4}$ b) $\frac{3}{4}$

Exercise 17d   p. 174

1. £12
2. 43 p
3. £140
4. £32.40
5. £81
6. £15
7. 6 p
8. £174

9. £63
10. 10 minutes
11. 60 cm
12. 12 kg
13. 1.08 litres
14. 4 km
15. 1.44 m
16. 1232 yards

17. 3 miles
18. 36 ft
19. 130
20. 52
21. a) 38%   b) 1440
22. 133
23. a) 1200   b) 800
24. 450

Exercise 17e   p. 175

1. $\frac{1}{2}$
2. $\frac{1}{3}$
3. $\frac{1}{4}$
4. $\frac{3}{10}$
5. $\frac{2}{3}$
6. $\frac{2}{5}$
7. $\frac{1}{8}$

8. $\frac{9}{20}$
9. $\frac{3}{5}$
10. $\frac{7}{40}$
11. $\frac{1}{5}$
12. $\frac{3}{8}$
13. $\frac{3}{8}$
14. $\frac{21}{100}$

15. $\frac{1}{5}$
16. $\frac{1}{20}$
17. $\frac{2}{3}$
18. $\frac{1}{10}$
19. $\frac{2}{5}$
20. $\frac{3}{16}$

Exercise 17f   p. 176

1. 75%
2. 40%
3. 80%
4. 25%
5. 85%
6. $33\frac{1}{3}$%
7. 80%
8. 45%
9. 60%

10. 12%
11. 50%
12. 30%
13. 20%
14. 60%
15. 45%
16. 45%
17. 44%
18. 20%

19. 75%
20. 80%
21. 20%
22. 62%
23. 72%
24. 25%
25. 84%
26. 35%

Exercise 17g   p. 177

1. 60%
2. a) 80%   b) 20%
3. a) 35%   b) 65%
4. 20%
5. a) 20%   b) £68.80
6. 14%

7. 12%
8. 6%, 94%
9. a) 50%   b) 50%
10. a) 64%   b) 36%
11. a) 50%   b) 50%
12. a) $33\frac{1}{3}$%   b) $66\frac{2}{3}$%

13. a) $31\frac{1}{4}$%   b) $68\frac{3}{4}$%
14. a) 25%   b) 75%
15. a) 36%   b) 64%
16. a) 25%   b) 75%
17. a) 44.4% ($44\frac{4}{9}$)   b) 55.6% ($55\frac{5}{9}$)
18. a) 25%   b) 75%

Exercise 17h   p. 179

1. £9
2. £50, £450
3. £36
4. £15

5. £2240
6. £24
7. 20%
8. 15%

9. $7\frac{1}{2}$% or 7.5%
10. 40%

Exercise 17i   p. 180

1. a) 0.8
 b) $\frac{4}{5}$

2. £180
3. 35%

4. 8%
5. £28

Exercise 17j   p. 180

1. a) 40%
 b) $\frac{2}{5}$

2. 20 minutes
3. 15%

4. 30%
5. 20%

Exercise 17k   p. 181

1. a) 0.3   b) 30%
2. $\frac{7}{20}$

3. 72 cm
4. 25%

5. £63
6. a) £19.80   b) 55%

Exercise 17l   p. 181

1. a) 0.55
 b) $\frac{11}{20}$

2. 32 cm
3. 20%

4. £11, £44
5. £9000

CHAPTER 18   Earning Money

Most pupils in their fifth year will be looking for paid work fairly soon, so the relevance of this chapter should be obvious. Linking with job advertisements and a discussion of which types of job offer which form of payment will add to the interest on this topic.

Exercise 18a   p. 182

1. £80
2. £114
3. £113.75

4. £192.40
5. £3
6. £3.50

7. £4.20
8. £4.25

Exercise 18b   p. 184

Up-to-date information, available from Job Centres and national and local newspapers, can be used for further questions based on this exercise and on subsequent exercises in this chapter.

1. a) £6 per hour
 b) £4.50 per hour
2. a) £8 per hour
 b) £5 per hour
3. a) £6.60 per hour
 b) £5.50 per hour
4. £6.36

5. £192
6. £212
7. £201.60
8. £2.60, £3.25, £22.75
9. a) £94.50   b) £4.05
 c) £32.40   d) £126.90

10. a) 7.00 a.m.
 b) 3.30 p.m.
 c) i) 5 hours   ii) $2\frac{1}{2}$ hours
 d) $37\frac{1}{2}$ hours
 e) Yes, on Friday
 f) £112.50

Exercise 18c   p. 186

1. a) £8   c) £15
 b) £24   d) £45
2. a) £14   b) £120
 c) £175

3. a) £200   b) £170
4. a) £145   b) £145
 c) £95
5. £150

Exercise 18d   p. 187

1. £17
2. a) £12   b) 10
   c) £16
3. a) £52   b) £97
4. a) 298   b) 248
   c) 50   d) £62.10
5. a) Tracy Cummings 112,
      Trevor Cross 101,
      Johanna Craig 96,
      Kim Chang 122

   b) Monday
   c) Kim Chang
   d) Trevor Cross

| | e) Number of kettles | | f) Income for |
	(i) at 60 p	(ii) at 75 p	the week
Tracey Cummings	75	37	£72.75
Trevor Cross	60	41	£66.75
Johanna Craig	75	21	£60.75
Kim Chang	75	47	£80.25

Exercise 18e   p. 188

It is interesting to compare weekly, monthly and annual earnings: mental approximations should be encouraged, e.g. £100 per week is about £5000 p.a., £12000 p.a. is about £240 per week or £1000 per month.

1. £10608
2. £3744
3. £5460
4. £12636
5. £364
6. £572
7. £228
8. £300
9. Fred, by £1060 a year
10. a) £9504
    b) £9620
    c) Elaine, by £116 a year

Exercise 18f   p. 190

Understanding percentage pay increases is important.

1. a) £9   b) £99
2. a) £4.20   b) £74.20
3. a) £6.05   b) £211.75
4. 8% – it is 8p more than £10
5. £7 – it is 68p more than 4%

Exercise 18g   p. 191

Deductions amount to such a large fraction of most wage-earners' pay that estimating 'take-home' pay, or the actual amount received for overtime, is worth while. For many people, take-home pay is about two-thirds of their gross pay.

1. £12
2. £7.74
3. £13.12
4. £17.50
5. £10.37

Exercise 18h   p. 192

1. a) £12300   b) £3075
2. a) £92   b) £27.60
3. a) £10200   b) £2040
4. a) £205   b) £57.40
5. £18.09
6. £11.70
7. £191.40
8. £2125
9. £4055.52

CHAPTER 19   Savings

The Post Office and the building societies have several leaflets available giving valuable information on the various forms of savings they offer. Discussions on the merits of Premium Bonds, Savings Certificates, Post Office savings accounts, building society accounts and so on can generate more interest in this topic. For the section on banking, it is worth providing blank copies of cheques and pay-in slips to be filled in by the pupils. (Point out that many of them will find that they must have a bank account when they start work to have their wages paid into.)

Exercise 19a   p. 195

1. £60
2. £72
3. £100
4. £1122
5. £210
6. a) £45   b) £22.50
   c) £11.25
7. a) 7%   b) £84
   c) £1284
8. a) 4   b) 103

Exercise 19b   p. 196

1. a) £100.91   b) £368.99
   c) £236.11   d) £333.99
2. a) June 30   b) May 25
3. £11.40
4. a) £26.52   b) £135.20
5. Once a month
6. Yes. Five days.

7. a) £121.89   b) £13.31
8. a) £240   b) £2880
9. a) £1700.31   b) £1033.99
10. £10000
11. £241.94
12. a) £169.25   b) £72.69
13. a) £135   b) £34.25

14. a) 3   b) 9
15. a) 11   b) 41
16. 5p, 10p
17. 5
18. T. Stanley
19. E. M. Parslow
20. 135

CHAPTER 20   Special Triangles and Quadrilaterals

If the basic work on geometry was covered some time ago, then reminders about angles in triangles, quadrilaterals and symmetry are essential.

Exercise 20a   p. 199

1. Isosceles
2. Neither
3. Equilateral

4. Isosceles
5. Neither
6. Neither

7. Isosceles
8. Equilateral

Exercise 20b   p. 200

1. $66°$
2. $75°$
3. $40°$
4. $64°$

5. $58°$
6. $60°$
7. $26°$
8. $140°$

9. $70°$
10. $45°$
11. $40°$
12. $20°$

Exercise 20c   p. 202

1. a) All sides are 4 cm.
   b) All angles are $90°$
   c) AB, BC, CD and DA are all equal.
   d) AB and DC are parallel, AD and BC are parallel.
   e) Square

2. a) AB and DC are 7 cm, AD and BC are 3.5 cm. All angles are $90°$
   b) AB = DC and AD = BC
   c) AB ∥ DC and AD ∥ BC (∥ means 'is parallel to')
   d) $\hat{A} = \hat{B} = \hat{C} = \hat{D}$
   e) Rectangle

3. a) PQ and SR are 5 cm, PS and QR are 9 cm. $\hat{P} = \hat{R} = 60°, \hat{S} = \hat{Q} = 120°$
   b) PQ = SR and PS = QR
   c) PQ ∥ SR and PS ∥ QR
   d) $\hat{P} = \hat{R}$ and $\hat{S} = \hat{Q}$
   e) Parallelogram

Exercise 20d   p. 205

1. a) All sides 6 cm, $\hat{D} = \hat{B} = 45°$, $\hat{A} = \hat{C} = 135°$
   b) AB = BC = CD = DA
   c) AB ∥ DC and AD ∥ BC
   d) Rhombus

2. a) PS = PQ = 7 cm, SR = RQ = 4 cm, $\hat{P} = 45°, \hat{R} = 85°$, $\hat{S} = \hat{Q} = 115°$
   b) PS = PQ and SR = RQ
   c) No
   d) $\hat{S} = \hat{Q}$
   e) Kite

3. a) No
   b) No
   c) ZY ∥ WX
   d) Trapezium

Exercise 20e   p. 206

1. Rectangle
2. Kite
3. Parallelogram
4. General quadrilateral

5. Trapezium
6. Rhombus
7. Square
8. General quadrilateral

9. Kite
10. Trapezium
11. Parallelogram
12. Rectangle

Exercise 20f   p. 207

1. $x = 120°, y = 45°$
2. $x = 118°, y = 62°$

3. $x = 106°, y = 110°$
4. $x = 75°, y = 105°$

5. $x = 105°$
6. $x = 112°, y = 68°$

Exercise 20g   p. 208

1. $x = 70°, y = z = 110°$
2. $x = 125°, y = 55°, z = 35°$

3. $x = 70°, y = 70°, z = 110°$
4. $x = 135°, y = 45°$

5. $x = 90°, y = 45°$
6. $x = 120°, y = 90°$

Exercise 20h   p. 209

Pupils can be asked to find examples of other patterns using the shapes discussed in this chapter. (Wallpaper and tiling patterns are a good source.)

1. A: trapezium,
B: isosceles triangle,
C: trapezium,
D: square,
E: isosceles triangle
2. $x = 75°, y = 105°$
3. $x = 60°, y = 120°, z = 120°$
4. $x = 76°$

5. a) Parallelogram, trapezium
and square
b) $x = 135°, y = 135°$
c) 120 cm
d) Rotational symmetry of
order 4
6. $x = 135°, y = 90°$
7. Equilateral triangle and
rhombus

8. a) $x = 60°, z = 60°, y = 120°$
b) 60°
c) All sides are 2 cm long.
9. a) ABCD and AECF
b) i) 4 cm   ii) 4 cm
iii) 4 cm   iv) 4 cm

CHAPTER 21   Transformations

A project on this work could be to give the pupils a basic shape and ask them to design a pattern using transformations of the shape. If they have seen the copy masters for Exercise 21f, they will have some idea of what is possible.

Diagrams in this chapter are not drawn full size.

Exercise 21a   p. 213

1.

2.
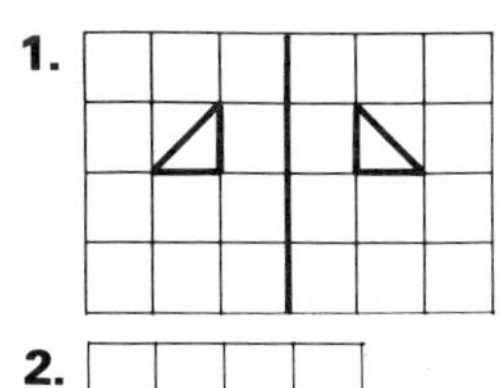

3.
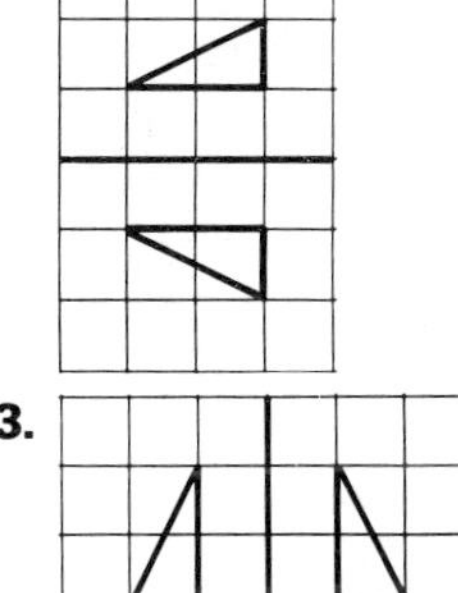

4.
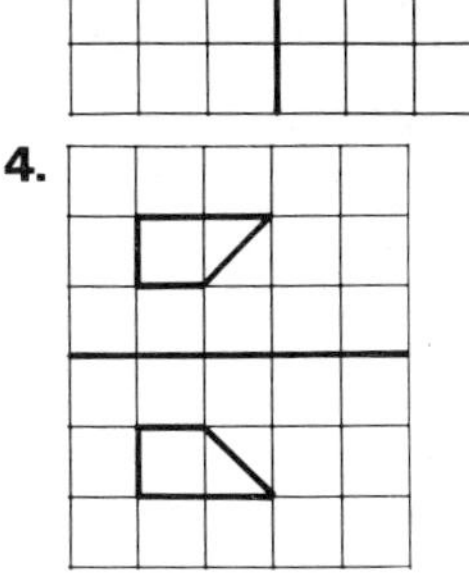

5.

6.
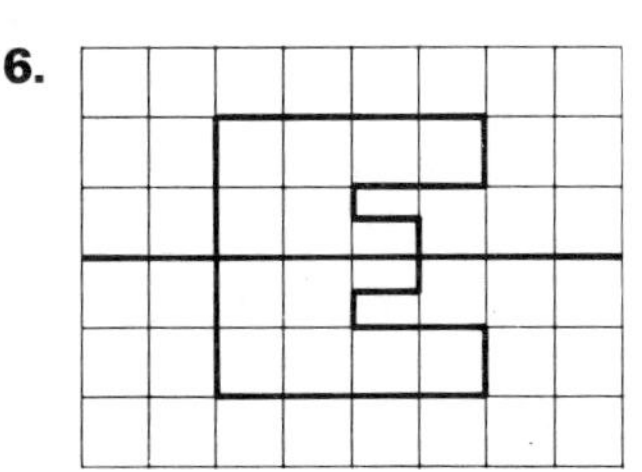

7.
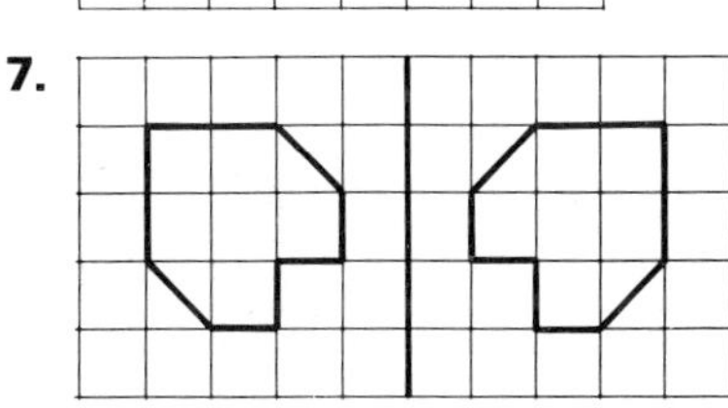

8.
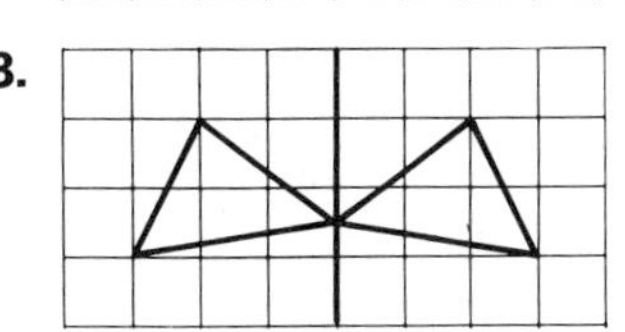

Exercise 21b   p. 215

1. 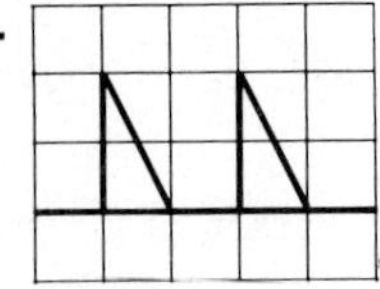

2. 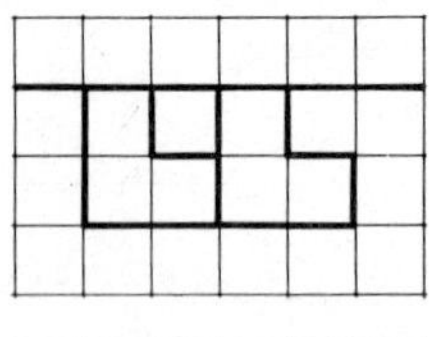

3. 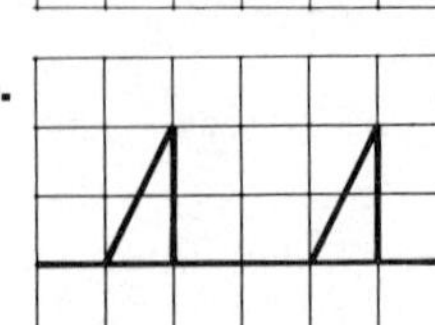

4. 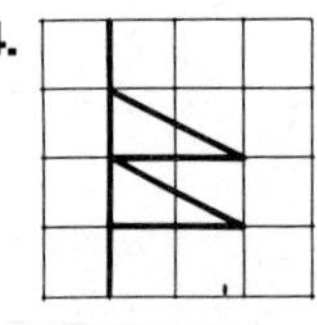

5. 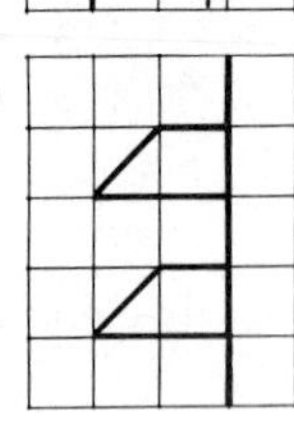

6. 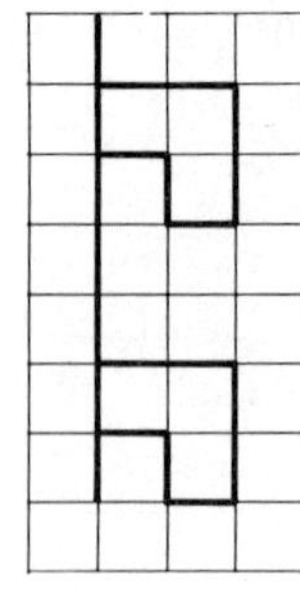

7. 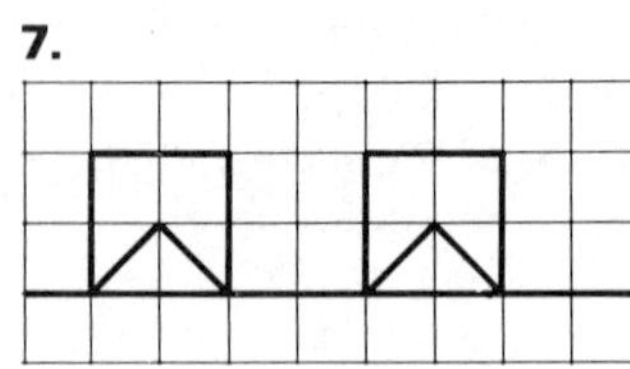

8. 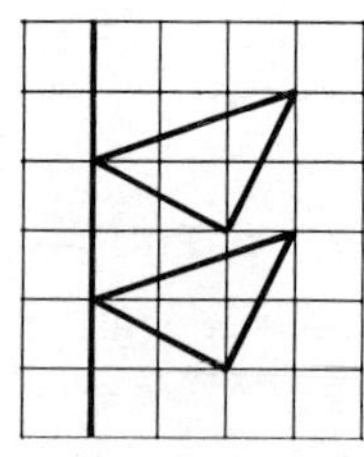

9. 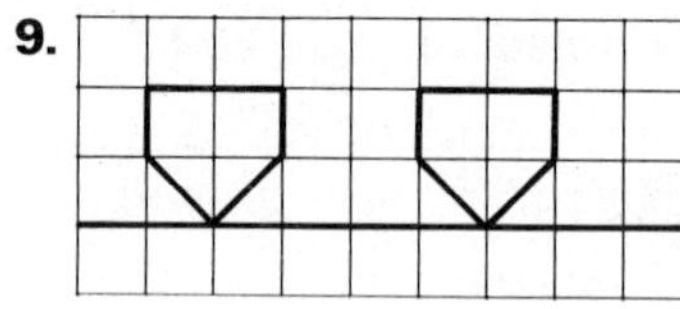

Exercise 21c   p. 217

Questions 1(c), (d), 2(a) and 3(a) involve knowledge of coordinates. If this chapter is worked before Chapter 22, it may be sensible to omit these questions.

1. a) A′   b) D′
    c) (3, 5)   d) (6, 4)

2. a) (6, 2)
    b) 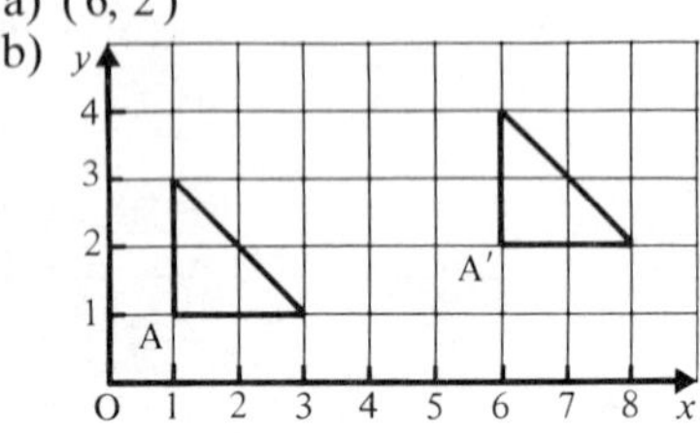

3. a) (1, 3)
    b) 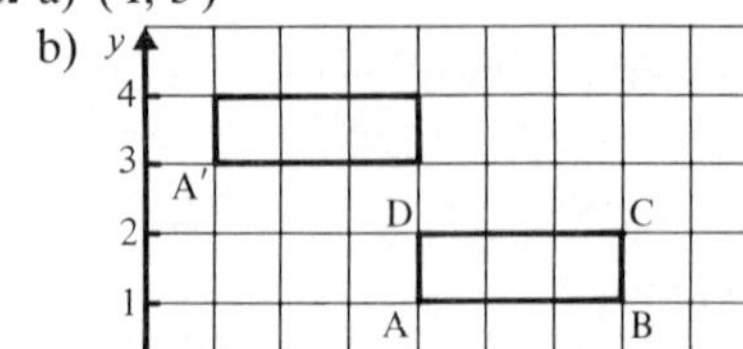

4. a) (5, 2)
    b) 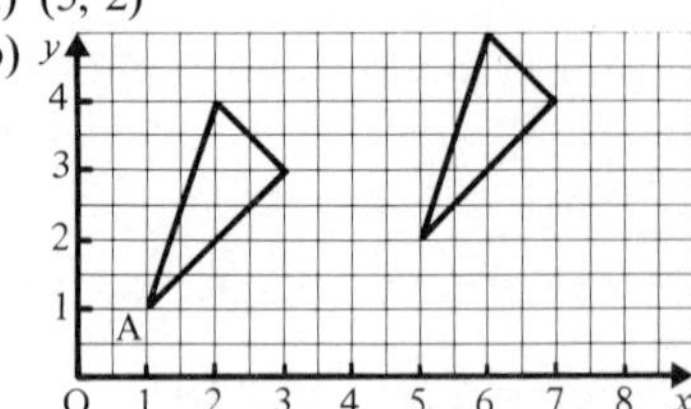

Exercise 21d   p. 219

1. 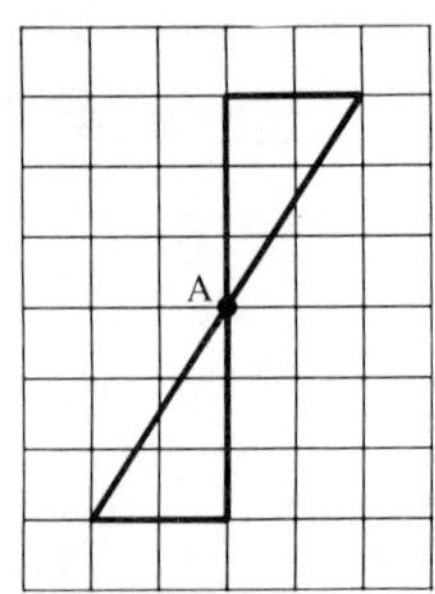

2. 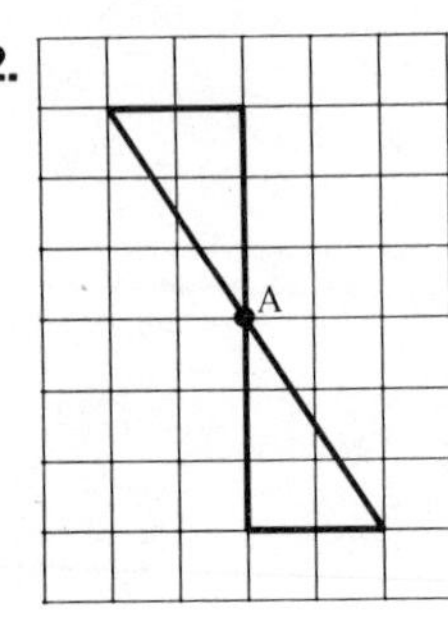

3. 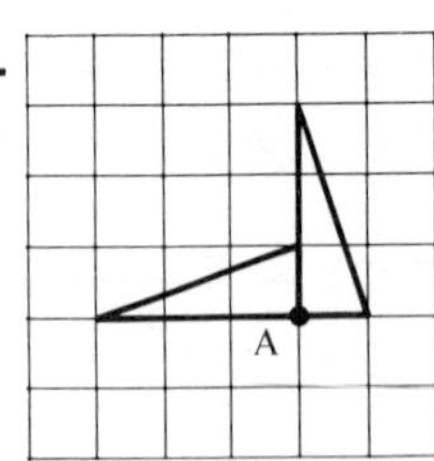

4. 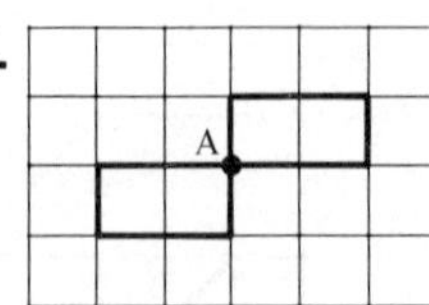

5. 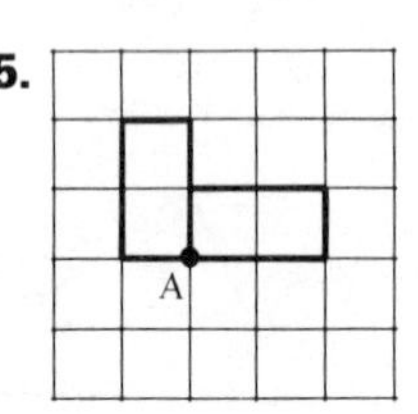

6. 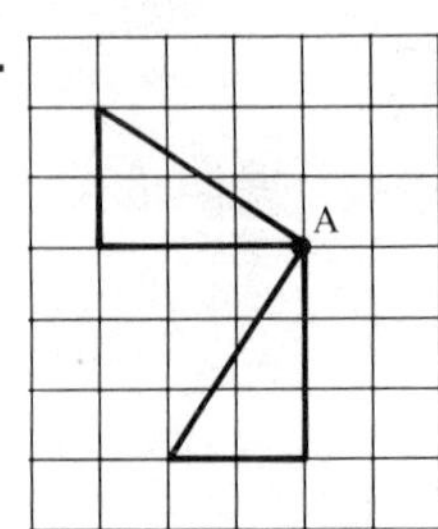

7.

8. 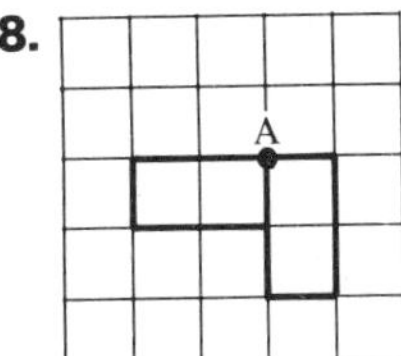

9. 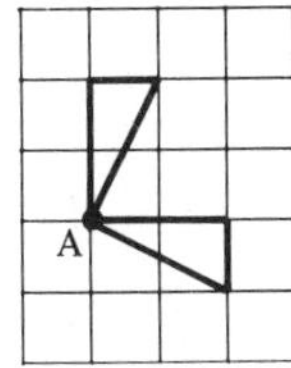

10. 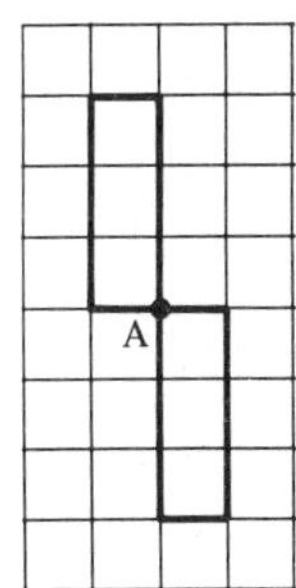

Exercise 21e   p. 222

1. 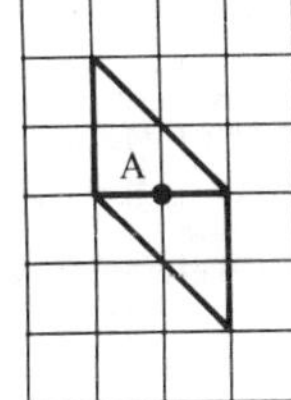

2. 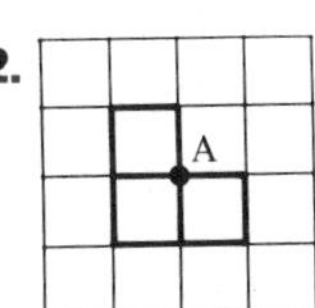

3. 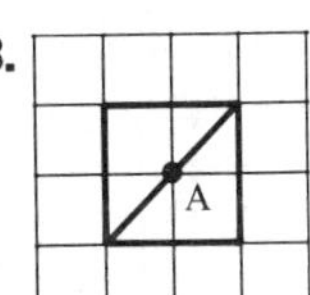

4. 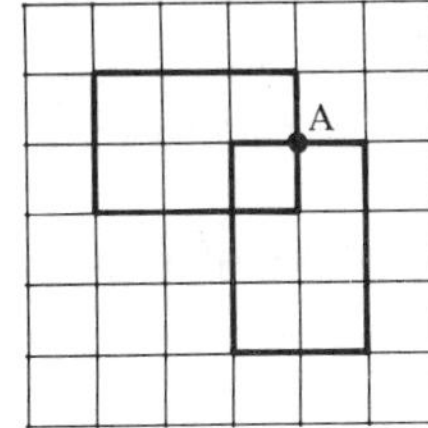

5. 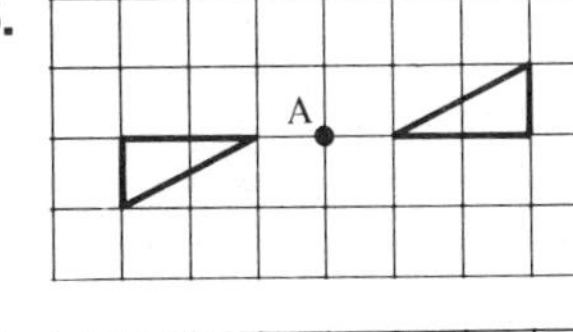

6. 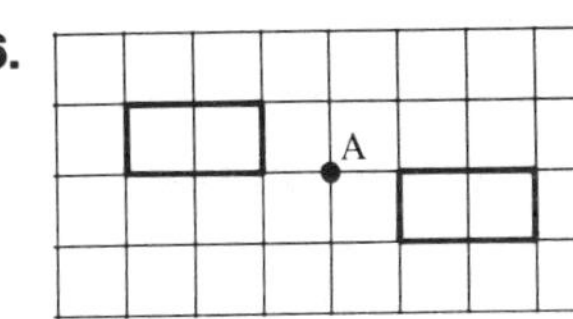

7. 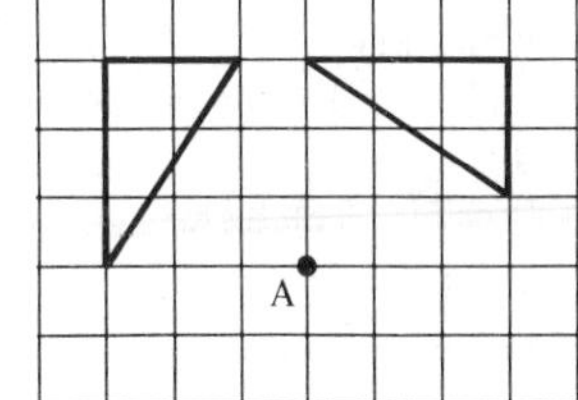

8. 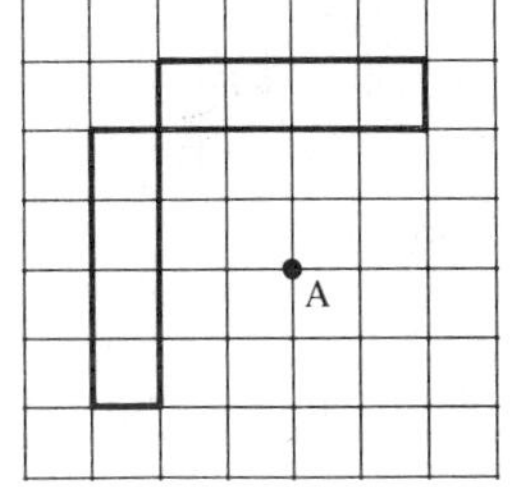

CHAPTER 22   Coordinates

If access to a computer is possible, a little knowledge of the DRAW and PLOT codes in BBC BASIC enables various figures to be drawn on the screen. This brings coordinates to life and helps tie up this topic with the last chapter.

Exercise 22a   p. 225

1. a) A(2, 7), B(8, 7), C(8, 3)
   b) (5, 7)
   c) (5, 5)
   d) (2, 3)

2. 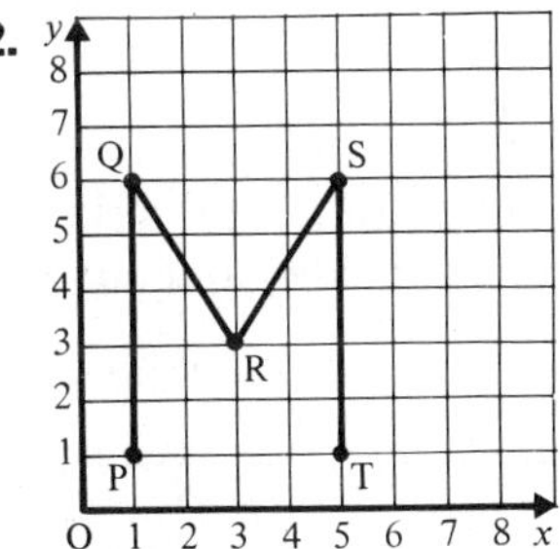
   b) M
   c) One line of symmetry
      through R parallel to PQ

3. a) N(1, 1), T(5, 3), Q(7, 7), R(10, 4)
   b) 8.5 cm on map, i.e. 85 km
   c) P(4, 4)

4. 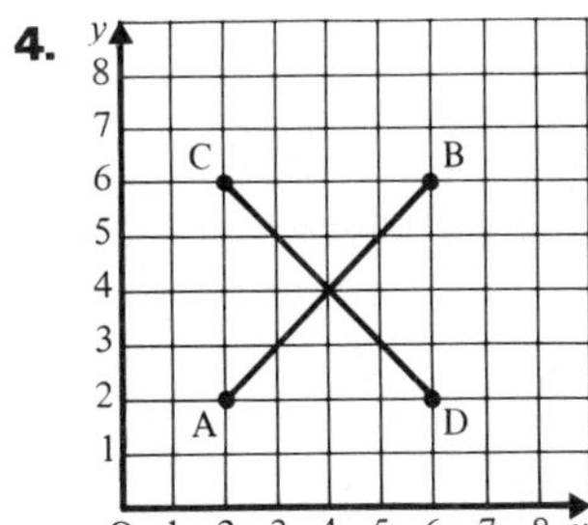
   b) X
   c) 4

5. 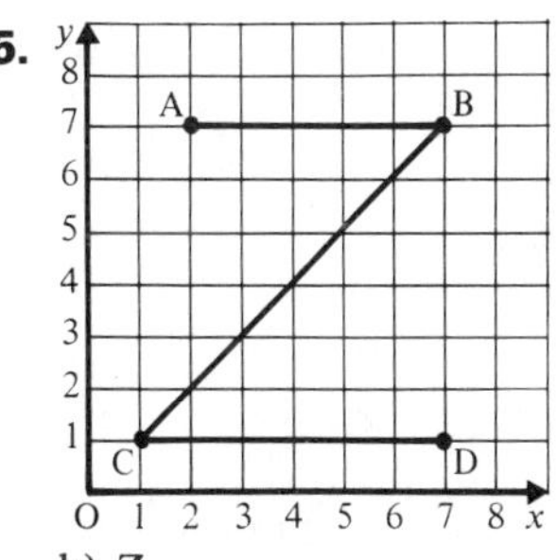
   b) Z
   c) 2

6. a) P(1, 1), Q(9, 7), R(3, 8)
   b) i) 7.3 cm   ii) 6.1 cm
   c) 96°

7. 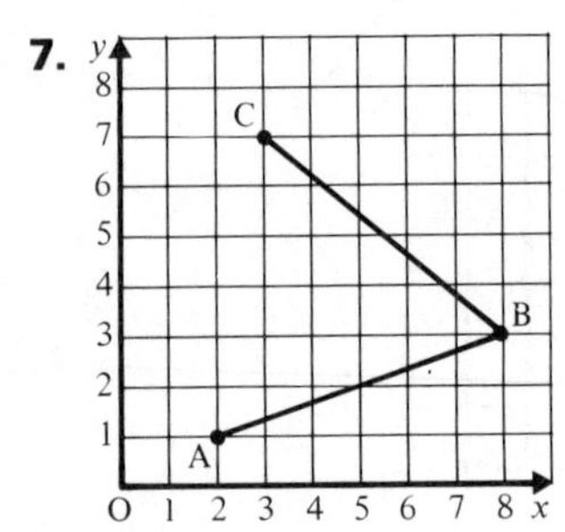

b) 6.3 cm   c) 6.4 cm

8. 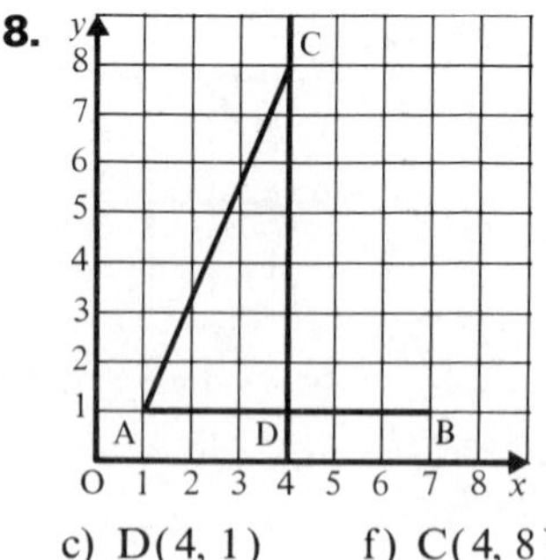

c) D(4, 1)   f) C(4, 8)   g) 67°

Exercise 22b   p. 228

1. a) A(1, 6), B(1, 2), C(5, 2)
  b) Isosceles and right-angled
  c) (3, 4)
  d) 13.7 cm   e) 8 cm²

2. 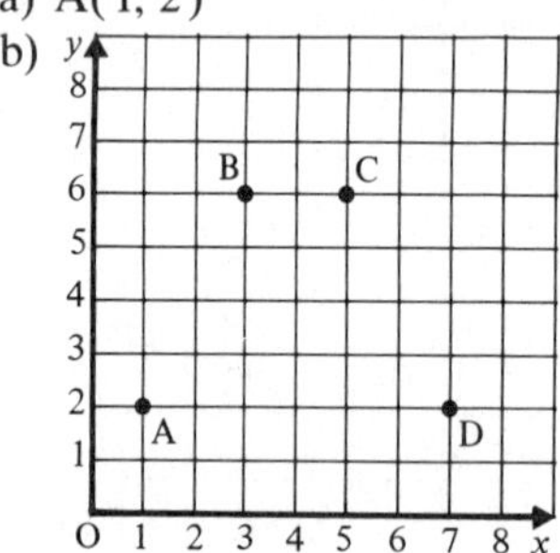

b) Parallelogram
c) 8 cm²

3. 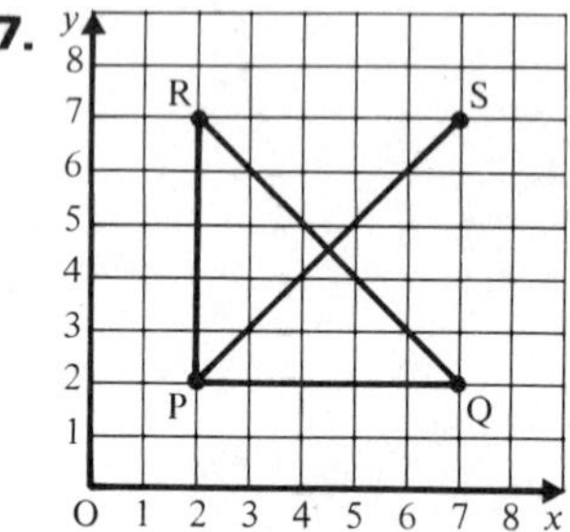

b) 12 cm
d) The square can be in other
   positions, provided that one
   corner is at A and the sides
   are 3 cm.

4. a) A(1, 1), B(6, 1), C(6, 4)
  b) D(1, 4)   c) 15 cm²

5. a) A(1, 2)
  b) 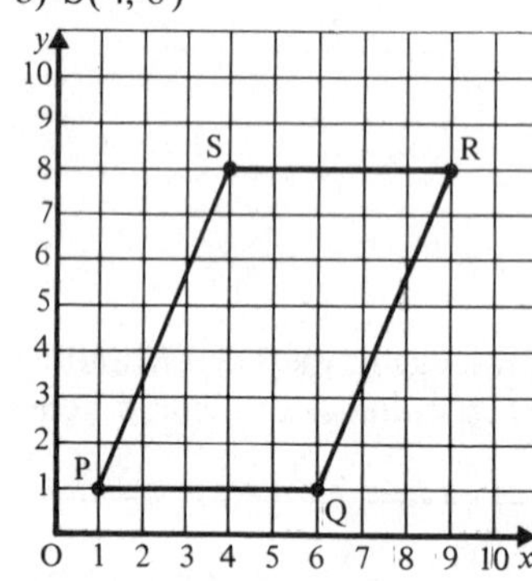

c) Trapezium
d) One line of symmetry

6. a) P(1, 1), Q(6, 1), R(9, 8)
  b) S(4, 8)

c) PR = 10.6 cm, SQ = 7.3 cm

7. 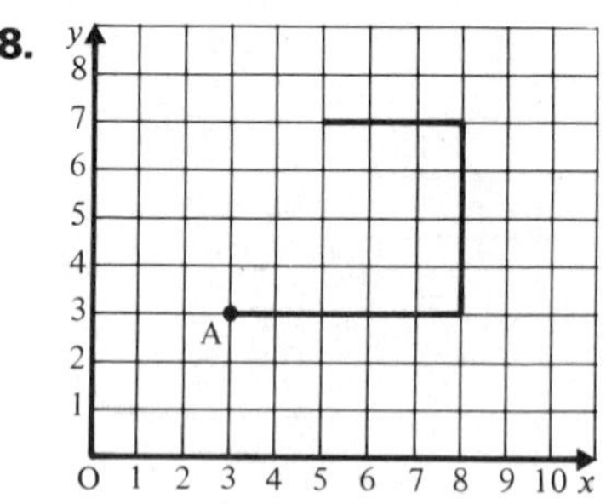

c) Isosceles triangle
e) (4.5, 4.5)

8. 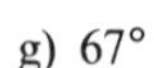

a) A(3, 3)
c) (3, 3). Trapezium.
d) (8, 7)

CHAPTER 23   Graphs

Provide some graphs from newspapers etc. so that you can discuss those that present information fairly and those that do not.

Exercise 23a   p. 232

1. a) i) 6.5 m   ii) 8.5 years
  b) i) 5 m   ii) 7 years
  c) Its 4th year

2. a) 78   b) 1 min
  c) 85   d) 93.8
  e) After 3.9 min. She stopped
     pedalling.

3. a) 20 ft   b) 36 ft
  c) 2 s

4. a) 5.4 m
  b) 7 a.m. and 11 a.m.

Exercise 23b   p. 235

1. a) £21.40
b) $56
c) £35.70 (≈ £36)
d) $14

2. a) 3.5 oz butter
3.5 oz sugar
3.5 oz chocolate
3 eggs
1.75 oz ground almonds
1.75 oz flour
b) Yes; 26 p

3. a) 14°F
b) 37.6°C (≈ 38°C)
c) 26.4°C (≈ 26°C)
d) 50°F

Exercise 23c   p. 238

1.
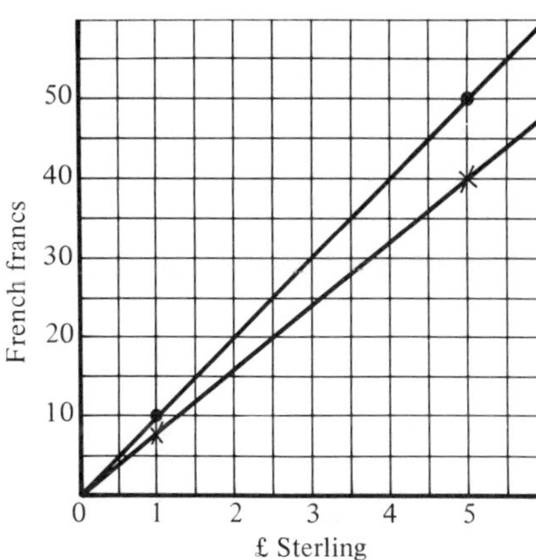

3. £5 = 50 Ff
6. a) 45 Ff   b) £2.50

Exercise 23d   p. 241

1. a) $3\frac{1}{2}$ h   c) $\frac{1}{2}$ h
b) 110 miles   d) 90 miles

2. a) 46 miles   e) 12.20
b) 2 min   f) 45 min
c) 20 miles   g) 9 miles
d) 12.18

3. a) 16 miles   d) 13.45
b) $\frac{1}{2}$ h   e) $8\frac{1}{2}$ miles, 15 min
c) $1\frac{1}{2}$ h

Exercise 23e   p. 244

1. a) 3 miles   c) $\frac{1}{2}$ h
b) 30 min   d) 6 m.p.h.

2. a) 86 miles   d) 36 miles
b) 2 h   e) 45 min
c) 43 m.p.h.   f) 48 m.p.h.

3. a) 800 miles   d) $1\frac{1}{2}$ h
b) 2 h   e) $2\frac{1}{2}$ h
c) 400 m.p.h.   f) 320 m.p.h.

Exercise 23f   p. 248

Graphs are drawn half size.

1.
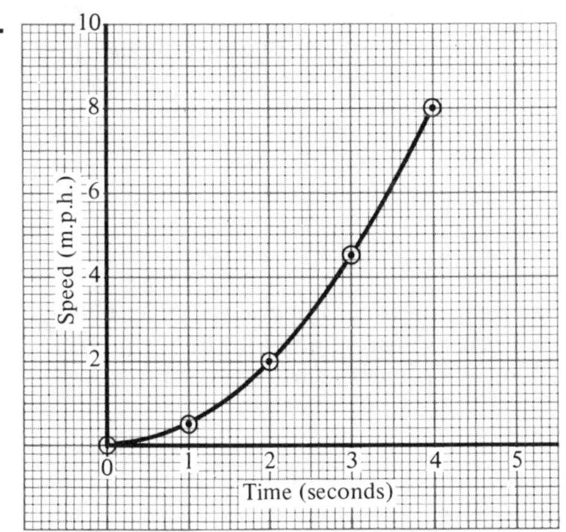

2.
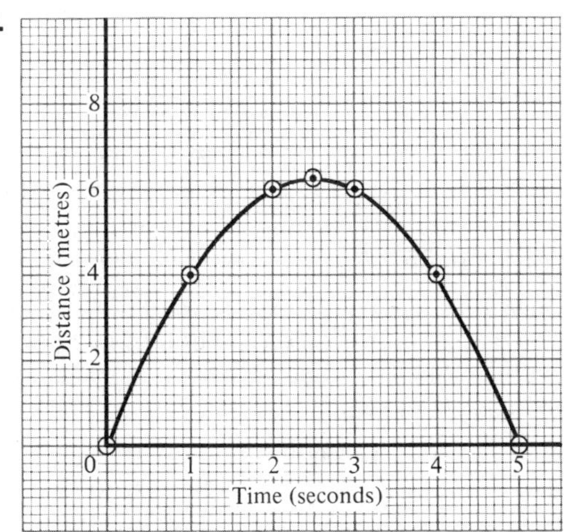

3.

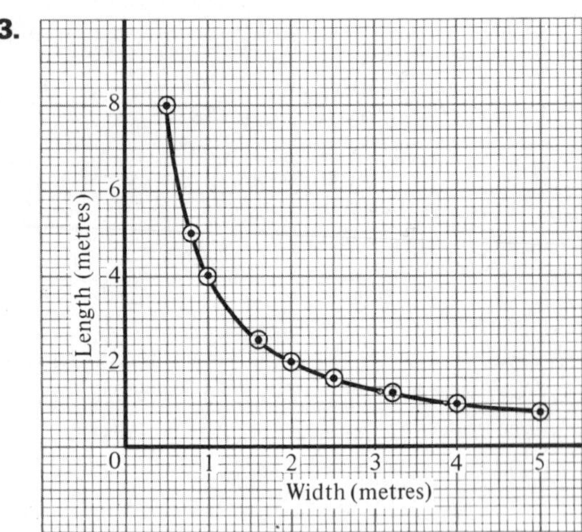

CHAPTER 24   Spending Money

Exercise 24a   p. 250

The information at the start of this exercise can be used as a source for aural questions.

1. 76 p	**9.** 36 p	**14.** £2.20
2. 15 p	**10.** 50 p	**15.** 13, 25 p
3. 75 p	**11.** a) Five 10 lb packs	**16.** 166
4. 96 p	b) £7.25	**17.** £174
5. 81 p	**12.** a) One 10 lb pack	**18.** a) £2.88
6. 42 p	b) £1.45	b) £5.74
7. 56 p	**13.** a) Two 56 lb bags	c) £4.26
8. 48 p	b) £15	**19.** £59 profit

Exercise 24b   p. 252

1. £260	**7.** £222.40	**13.** a) £135.40 b) £11.40
2. £530	**8.** a) £206 b) £31	**14.** a) £825 b) £3435
3. £2760	**9.** a) £930 b) £230	c) £685
4. £201.50	**10.** a) £470.40 b) £110.40	**15.** a) £50 b) £302
5. £460	**11.** a) £7368 b) £1068	c) £52
6. £347.40	**12.** a) £258.13 b) £38.14	

Exercise 24c   p. 256

When this exercise has been worked, other forms of credit can be discussed such as bank loans, personal loans from finance houses, etc. A project on this topic could be based around collecting information on various ways of paying for a purchase such as a car, including the true cost of the purchase. It is worth pointing out that, although there is a good deal of advertising for loans etc., they are not automatically available: some form of credit rating and/or security is often required.

1. £5	**2.** Yes, the total is £799.20 but her credit limit is £800.	**3.** a) £42.50 b) £807.50

Exercise 24d   p. 257

Pupils are often surprised that the total cost of a house bought on a mortgage is about three times the initial purchase price. Point out that things are in no way as bad as they seem because (assuming inflation continues, even at a low rate) the value of the house increases considerably and the number of hours of work needed to pay the mortgage gets smaller and smaller as the years go by.

1. a) £480	b) £144 000	**4.** a) £310 b) £93 000	**7.** a) £57 000 b) £3000
2. a) £383.60	b) £138 096	**5.** £52 250	c) £684
3. a) £299	b) £822.25	**6.** a) £24 000 b) £6000	

Exercise 24e   p. 259

Stress how important it is to insure anything of value.

1. £71.50
2. £39
3. £54
4. £4.50

5. £61.10
6. £46.50
7. a) £500   b) £31.20
    c) £126.50

8. a) £37.80   b) £4.68
    c) £40.25

Exercise 24f   p. 260

1. £750
2. £700

3. £825
4. £1200

5. £1125
6. £495

CHAPTER 25   Electricity and Gas

This chapter deals with reading electricity and gas meters and the bills that follow. We have included a few dial meters but not many, as these are becoming obsolete. The pupils can be asked to read their own meters at home.

Exercise 25a   p. 262

1. 32657
2. 72358
3. 52734
4. 547
5. 470

6. 352
7. 36754
8. 229
9. 28199
10. 19786

11. Second quarter 334, third quarter 192, fourth quarter 473; 1641
12. The missing values are 47920, 48314, 48841

Exercise 25b   p. 264

1. £64.14
2. £18.50
3. £40.23
4. a) i) 49263   ii) 50209
    b) 946
    c) 6.00 p
    d) £56.76
    e) £8.50   f) £65.26

5. 750 units, £85
6. 1200 units, £108
7. 1120 units, £84.30
8. 20 min
9. 2 h
10. 10 h
11. a) 24   b) £1.80

12. a) 60   b) £4.50
13. a) 10   b) 75 p
14. a) 45   b) £3.38

Exercise 25c   p. 268

1. 4143
2. 234

3. 317
4. 197

Exercise 25d   p. 269

1. a) 39 p   c) 289
    b) 0%   d) 300

2. a) 42 p   d) £118.53
    b) 274   e) £128.13
    c) 282.22

3. £130
4. £213
5. £264.73

CHAPTER 26   Telephone Bills and Postage

The information on charges in this chapter is correct at the time of writing (1988) but will be out of date fairly quickly. For extra work, which could be aural, it is worth using the current charges. This information is freely available from British Telecom and the Post Office respectively.

Exercise 26a   p. 272

1. a) £20.15
    b) 528
    c) i) 016927   ii) 017455
    d) £49.19
    e) 15%
    f) £7.38
    g) £56.57

2. a) 3   f) 5.5 p
    b) £14.95   g) £46.37
    c) £7.50   h) £68.82
    d) i) 17471   i) 15%
        ii) 18314   j) £10.32
    e) 843   k) £79.14
3. £57.50

4. £52.90
5. £64.61
6. £88.59
7. £62.22
8. £87.63
9. £58.92
10. £63.39

Exercise 26b   p. 275

1. a) 3
   b) 12
   c) 15
   d) $4\frac{1}{2}$

   e) $7\frac{1}{2}$
   f) $10\frac{1}{2}$
   g) 9
   h) $22\frac{1}{2}$

2. a) 16
   b) 24
   c) 12
   d) 4

   e) 20
   f) $13\frac{1}{3}$
   g) 28
   h) 40

3. a) 4
   b) 11
   c) 7

4. a) 1
   b) 4
   c) 4

5. a) 2
   b) 18
   c) 11

6. 5 p
7. 70 p
8. £ 1.20
9. £ 3.20

Exercise 26c   p. 278

1. 72 p
2. 65 p
3. 75 p
4. 68 p

5. £ 1.02
6. £ 1.04
7. £ 3.80
8. £ 12.75

9. £ 7.05
10. £ 3.65

CHAPTER 27   Holidays

Most travel agents and tourist information offices supply vast quantities of free literature which can be used for a project on planning a holiday. Revise the work on timetables in Chapter 4: then the project can include the making of a complete itinerary, together with the total cost. For holidays abroad, up-to-date currency exchange rates are published daily by several newspapers.

Exercise 27a   p. 279

1. a) £ 196   b) £ 252
   c) £ 248
2. a) £ 284   b) £ 269

3. £ 90 each
4. £ 7
5. a) No   b) Yes up to 16 July

6. a) £ 132.50   b) £ 170
   c) £ 66.50

Exercise 27b   p. 281

1. a) £ 494   b) £ 110
2. a) £ 116   b) £ 274

3. a) £ 329   b) £ 164.50
   c) £ 987   d) £ 132

Exercise 27c   p. 282

1. a) 6000 pesetas   b) 34 000 pesetas
2. a) 1320 francs   b) 735 francs
3. a) £ 30   b) £ 183
4. a) £ 9   b) £ 82

5. a) 1260 francs   b) £ 31
6. a) 14 400 pesetas   b) £ 65
7. a) £ 320   b) $ 45
8. a) 158 400 lire   b) £ 45

9. a) 4480 francs   b) £ 36
10. a) 145 DM   b) £ 70

Exercise 27d   p. 283

1. a) $ 12   c) £ 10
   b) £ 4   d) $ 9
2. a) 90 Ff   c) 54 Ff
   b) £ 5   d) £ 9

3. a) 900 pta   c) 1530 pta
   b) £ 4.50   d) £ 5.55
   e) 1800 pta, 180 pta

Exercise 27e   p. 285

1. 874 francs
2. a) 9900 pesetas   b) £ 25
3. a) $ 150   b) £ 500

4. a) £ 645   b) £ 516
5. £ 700

CHAPTER 28   Statistics

Information which people need to understand is often presented in bar charts or other statistical forms. This information includes pie charts sent out with local government tax demands (still collected in the form of rates at the time of writing), a wealth of bar charts in the run-up to a general election, bar charts relating to business activities, and so on. Most people at work will be at an advantage if they can 'read' statistical diagrams. It is worth finding some examples of the diagrams mentioned above.

Exercise 28a   p. 287

1. a) Frequencies 5, 9, 9, 5, 7, 3, 2
   Total 40
   b) 5
   c) 9
   d) 5

2. a) Frequencies 4, 5, 6, 8, 6, 3
   Total 32
   b) 6
   c) 14
   d) $\frac{8}{32}$ i.e. $\frac{1}{4}$ e) 500

3. a) Frequencies 6, 8, 2, 4, 4
   Total 24
   b) It is a check that you have counted correctly.
   c) Blue   d) Purple   e) No

Exercise 28b   p. 289

1. a) 24
   b)

   c) 150 cm

2. a) 9
   b) The Two Connies
   d)

3. a)

   b) 3000
   c) 3500
   d) Yes

Exercise 28c   p. 291

1. a) 10

b) 5

c)
Vehicle	Bicycle	Van	Motorcycle	Car	Lorry
Number	5	15	10	25	5

2. a) 3

b) 2

3. a)
Year	1985	1986	1987	1988
Number of companies	21	72	56	49

b) 198

c) 1986

4. a) 5

b) 0

c)
Animal	Bird	Dog	Hamster	Snake	Cat
Number of animals	2	5	1	0	7

d) 15

e) No; some pupils might have no pet, or more than one pet.

5. a)
Fruit	Gooseberries	Blackcurrants	Raspberries	Strawberries	Redcurrants
Quantity picked (kg)	20	30	55	35	8

b) Raspberries

c) 148 kg

Exercise 28d   p. 293

1. a) 6

b) 14

c) 4

2. a) It is trying to shock people into taking more care.

b) The number of deaths might be even bigger.

3. a) 4

b) 4

c) 17

d) It is bigger.

e) It looks as if there are more rabbits than birds.

Each picture should take the same amount of space.

4. a) 50

b) 15

c)
Year	1880	1905	1930	1955	1980
Number of people	50	40	40	20	15

d) By drawing one-fifth of a pin-man.

It is difficult to tell how many people are represented by part of a figure.

Exercise 28e   p. 296

1. a) They are the same.
   b) $\frac{1}{4}$
   c) 60
2. a) Britain
   b) $\frac{1}{6}$
   c) $\frac{1}{4}$
   d) 25%
3. a) Red
   b)

Colour	Green	Yellow	Blue	Red
Angle	90°	60°	90°	120°

   c) $\frac{1}{4}$
   d) 6
   e) 4
   f) 8

CHAPTER 29   Averages

Most people think they know what 'average' means, but it is worth pointing out that the word is used fairly loosely by the media and any statement concerning averages should be thought about: is it mean, mode or median, and what exactly are they finding the average of ? Statements such as 'The average shoe size for women is 6', 'The average weekly wage is £150', 'The average teenager spends £5 a week on cassettes' can form the basis of a discussion. Collect such statements from newspapers etc.

Exercise 29a   p. 300

1. 7
2. 8
3. 33 g
4. 4 m
5. 12.5
6. 45
7. 6 mm
8. 72 min
9. $6\frac{1}{2}$ apples
10. a) 26   b) 25
   c) Giles' plants
   d) Malcolm's plants

Exercise 29b   p. 301

1. 2
2. 13
3. 80
4. 2
5. 5
6. 4
7. a) Size 8   b) 33
8. a) 5   b) 22

Exercise 29c   p. 303

1. 4
2. 21
3. 3.2 cm
4. 1420
5. 5 ft 6 in
6. 9 kg

Exercise 29d   p. 304

1. a) 48   b) 48
   c) No
2. a) 5.5   b) 5.2
   c) 5.4
3. 1.5
4. a) 71   b) 72
5. a) 83   b) 82
   c) The label applies to *all* the
      boxes of matches packed,
      not just the ten boxes checked.
6. a) 110   e) 103
   b) 172   f) The Record
   c) 36   g) The Record
   d) 412

Exercise 29e   p. 305

1. 7
2. 48
3. 4.5 cm
4. 10 kg
5. a) 4 cm
   b) 185 cm
   c) 29
6. a) 11   c) 11
   b) 11   d) 8
7. a) 8   c) None
   b) 7   d) 20
8. a) 19   d) 21
   b) 25   e) 127
   c) Wednesday
9. a) 10
   b) 70
   c) 60
   d) 5
   e) 250
   f) $\frac{1}{5}$

CHAPTER 30   Ratio and Proportion

Many pupils find this topic difficult, but will probably appreciate its use when it is applied to problems such as increasing quantities for recipes, decorating and so on. This chapter is too long to work through without several breaks. For example, Exercises 30a and 30b could be followed by the first two exercises in Chapter 31. Similarly Exercise 30c can be followed by Exercises 31c and 31d. Exercises 30d and 30e can be followed by more work from Chapter 31. Many of the problems in this chapter can be done intuitively, and this can be encouraged provided that pupils are also encouraged to look at their answer and ask themselves 'Is this a reasonable answer?' (They *should* do this for all their work, but getting them to do so is another matter.)

Exercise 30a   p. 308

1. $3:4$

2. $1:2$

3. $3:2$

4. $2:3$

5. $3:4$

6. $2:1$

7. $1:2$

8. $2:1$

9. $4:5$

10. $3:2$

11. a) $2:1$ b) $1:2$

12. a) $2:1$ b) $1:2$

13. a) $7:5$ b) $7:12$

Exercise 30b   p. 309

1. $8:3$

2. $1:4$

3. $1:72$

4. $10:9$

Exercise 30c   p. 310

1. a) 150 kg   b) 40 lb
   c) 4 bucketfuls

2. a) 18 kg   b) 48 kg

3. a) 4   b) 16
   c) 12 oz   d) 12 oz

4. 500 kg

5. a) 250 kg   b) 400 kg
   c) 750 kg

6. a) 960 g   b) 720 g
   c) 2280 g

Exercise 30d   p. 311

1. Annie 60 p, Phil 30 p

2. Les £ 15, Liz £ 25

3. Ann £ 32, Robin £ 24

4. June £ 135, John £ 90

5. 20 ft, 25 ft

6. 45, 10

7. £ 180

8. 8000 m², 3000 m², 5000 m²

9. 128

10. a) 610   b) 244

11. 120

12. a) 0.2 kg   b) 10 p

Exercise 30e   p. 313

1. 200 miles

2. a) £ 6   b) £ 24

3. a) 28 p   b) £ 1.40

4. a) 24 p   b) £ 1.20

5. 26 litres

6. a) £ 225   b) 14 m²

7. £ 2.16

8. £ 54.60

9. a) £ 4   b) 20

10. a) 225   b) 15

11. 8 miles

12. 42

Exercise 30f   p. 314

1. a) 36   b) 9
   c) 3

2. a) 24   b) 4

3. a) 200   b) £ 36
   c) 25

4. a) 144 miles   b) 3 h
   c) 24 m.p.h.

CHAPTER 31   Enlargement and Similarity

All drawings are given half size.

Exercise 31a   p. 316

1. 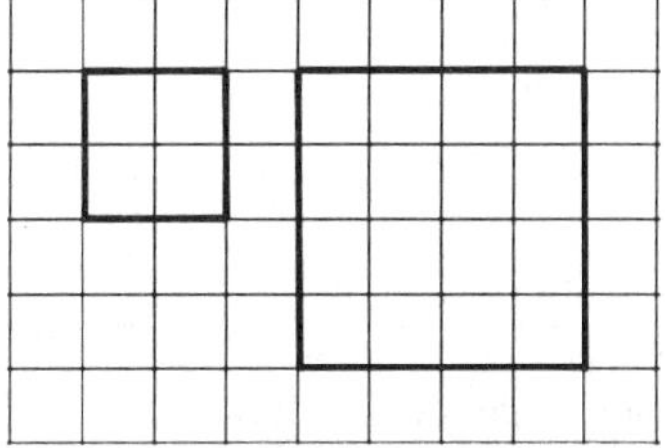

5. 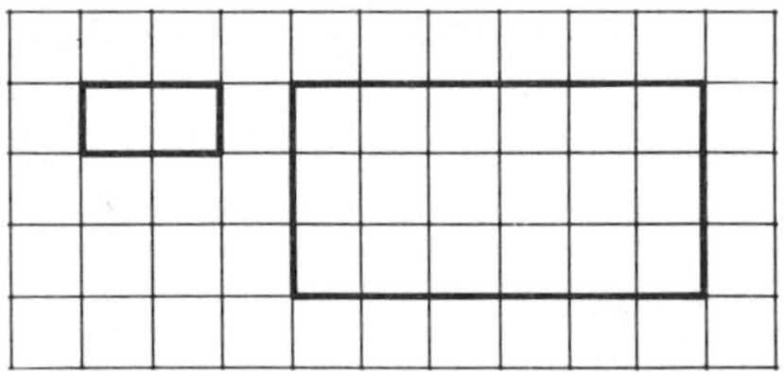

2. 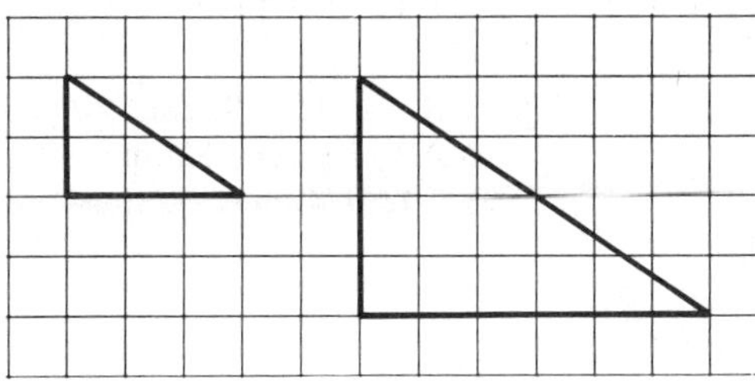

6. 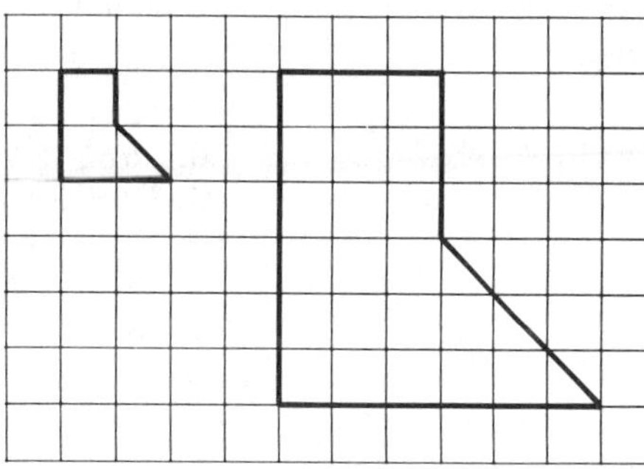

3. 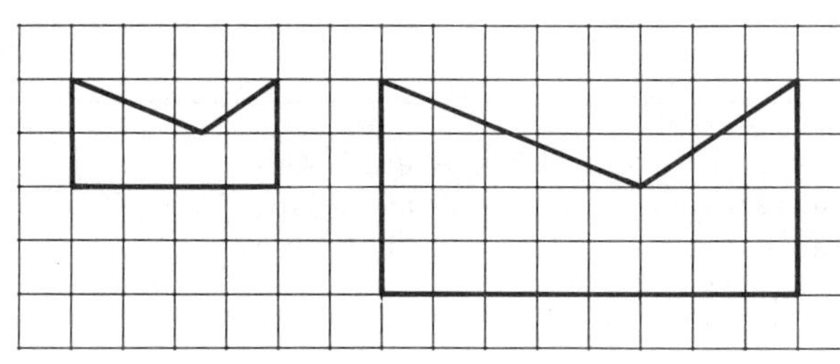

7. 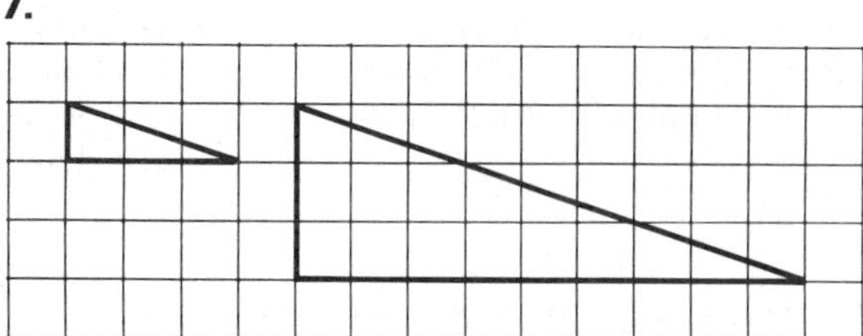

4. 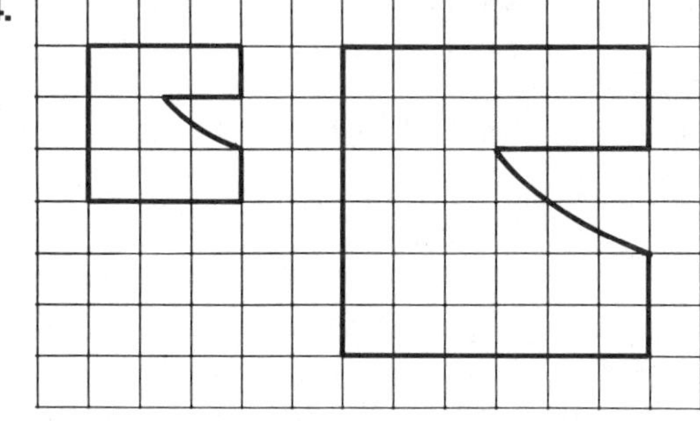

8. 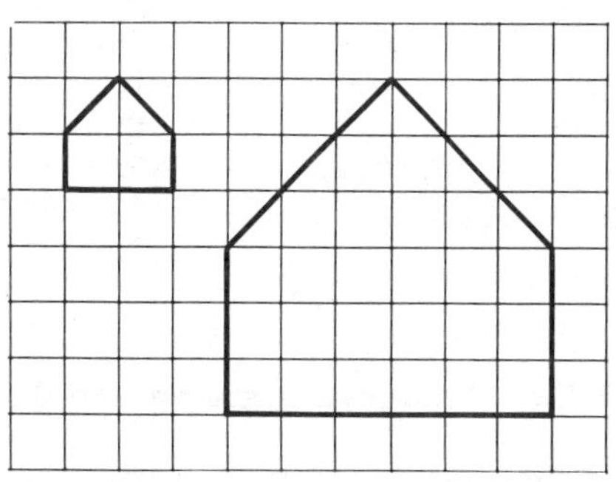

9. a) 6 cm   b) 3 cm   c) A circle with radius 6 cm   d) 12 cm

Exercise 31b   p. 318

1. 2   **3.** 4   **5.** 2
2. 3   **4.** 2

Exercise 31c   p. 319

1. Yes   **4.** Yes   **7.** Yes
2. No   **5.** Yes   **8.** No
3. Yes   **6.** No   **9.** Yes

Exercise 31d p. 321

1. a) 4   b) 20 cm
2. a) 4 cm   b) 16 cm
   c) 12 cm
3. a) 8 in by 12 in
   b) 24 sq in   b) 96 sq in
4. a) 20 cm   b) 8 cm
   c) 3 cm by 3 cm

5. a) Width = 4.5 cm
     Height = 3 cm
   b) Width = 45 cm
     Height = 30 cm
   c) 1.5 cm, 15 cm

6. a) 3 m
   b) 2 m
   c) 4.5 m
   d) i) 2 m^2 ii) 4.5 m^2

Exercise 31e p. 324

The pupils should have some real scale drawings to look at and interpret. Local estate agents may have some available. A scale drawing of your school should be available. The local authority might provide some for their public buildings.

1. 9 m
2. 6 m
3. 2 m
4. 4 m
5. 8 m^2
6. 32 m

7. 4.2 m
8. 3.6 m, 15.1 m^2 to 3 s.f.
9. 3.1 m, 2.3 m
10. No, the kitchen is only 340 cm long.
11. (Bedroom is 3.1 m × 3.6 m.), 10.8 m^2

12. a) BC = 9 cm, 90 cm
   b) AB = 3.2 cm, 32 cm
   c) 0.4 cm, 4 cm

Exercise 31f p. 326

1. a) 156 km   b) 228 km
2. a) 35 miles   c) 12 miles
   b) 20 miles   d) 24 miles

3. a) 750 m
   b) about 3 cm (measuring from the roundabout)
   c) i) 1500 m   ii) 1.5 km

4. a) about 430 m
   b) about 120 m
   c) about 70 m

Exercise 31g p. 328

1. a) 250 cm   b) 2.5 m
2. a) 2 m   b) 60 m
3. a) 1000 m   b) 5000 m
   c) 8000 m
4. a) 5 km   b) 30 km
5. 3.6 km

6. 10 km
7. a) 0.2 km   b) 10 km
8. a) 100 cm   b) 1 : 100
9. a) 100 000 cm   b) 1 : 100 000
10. a) 1000 cm   b) 1 : 1000
11. 1 : 5000

12. 1 : 10 000
13. 1 : 20 000
14. 1 : 250 000
15. 1 : 72
16. a) 1 : 200   b) 5 m

CHAPTER 32 Accurate Drawing and Scale Diagrams

It cannot be emphasised too often how important it is to produce neat tidy drawings done with a *sharp* pencil. Some revision on basic triangle and quadrilateral facts is needed for this chapter.

Exercise 32b p. 333

Answers should be within 1° of the answers given.

1. $\hat{P} = 22°$, $\hat{Q} = 50°$, $\hat{R} = 108°$
2. $\hat{Y} = 46°$, $\hat{X} = \hat{Z} = 67°$

3. $\hat{A} = 115°$, $\hat{B} = 43°$, $\hat{C} = 22°$
4. $\hat{L} = 83°$, $\hat{M} = 36°$, $\hat{N} = 61°$

5. $\hat{X} = 59°$, $\hat{Y} = 37°$, $\hat{Z} = 84°$
6. $\hat{A} = 80°$, $\hat{B} = 61°$, $\hat{C} = 39°$

Exercise 32c p. 334

5. b) 118°: your answer should be within 1° of this.
   c) The two angles at A should add up to 180°, but this accuracy is not always possible from a drawing.
     An answer from 178° to 182° is reasonable, but if your answer is outside this range, you have made a mistake.

Exercise 32d   p. 336

Answers should be within 0.1 cm of these answers.

1. 8.4 cm

2. 12.2 cm

Answers should be within 1° of these answers.

3. 54°
4. 104°

5. 90°
6. 11.1 cm

Exercise 32e   p. 337

1. b) 85°
2. b) 40°

3. a) 35°
4. a) 41°

Exercise 32f   p. 338

Angles should be within 1° of those given and lengths should be within 1 mm (i.e. 0.1 cm) of those given.

1. b) 114°   c) Parallelogram
2. b) 91°
3. e) 4.7 cm

4. b) 75°
5. a) Equilateral   b) 60°, 60°

Exercise 32g   p. 340

1. 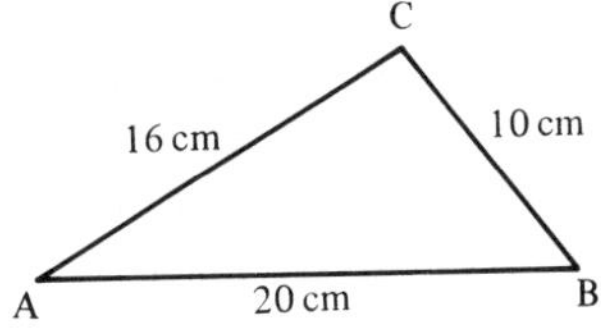

2. 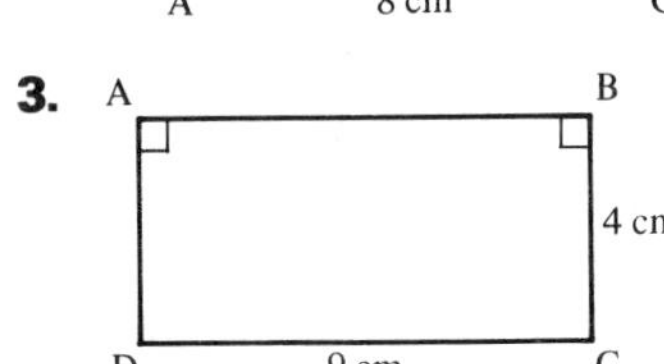

3.

4. 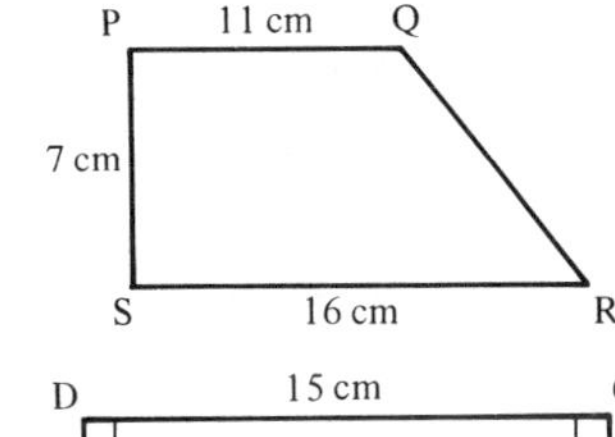

5.

6. 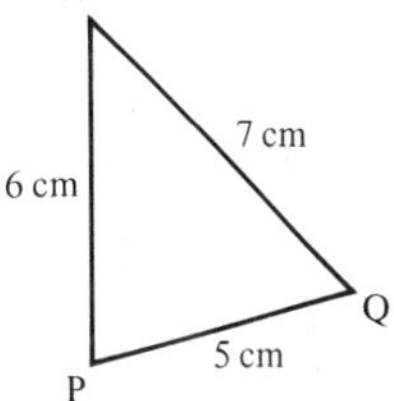

c) 44° (43°, 44° or 45° are acceptable answers.)
7. b) 111 km (110 km, 111 km, 112 km are acceptable answers.)

CHAPTER 33   Bearings

The first part of this chapter can be done quite early on in the two year course, but it will need revising before the harder scale drawings are tackled at the end of this chapter. (Note that not all syllabuses require bearings in the form S 20°W.)

Exercise 33a   p. 344

1. E
2. S
3. NW

4. 90°
5. 45°

6. Trees
7. Church

Exercise 33b   p. 345

1. 039°
2. 125°
3. 305°
4. 212°
5. 230°
6. 195°
7. 316°
8. 150°
9. a) 050°   b) 295°
   c) 265°
10. 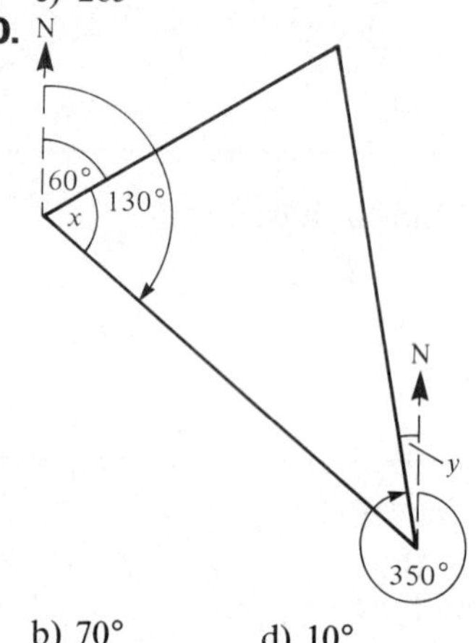
    b) 70°   d) 10°

Exercise 33c   p. 349

1. 055°
2. 123°
3. 092°
4. 060°
5. 242°
6. 297°
7. 233°
8. 324°
9.
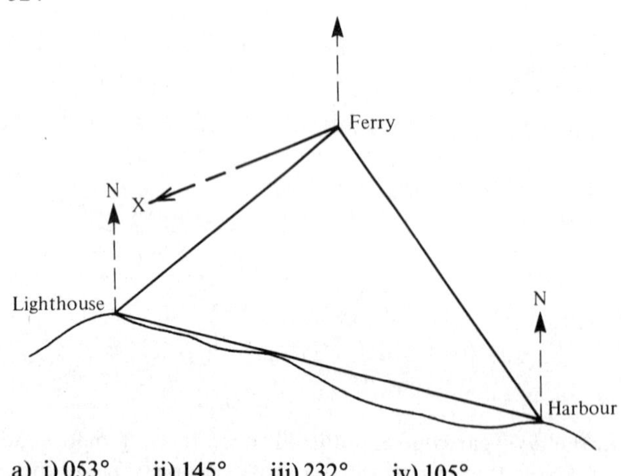

   a) i) 053°   ii) 145°   iii) 232°   iv) 105°

Exercise 33d   p. 352

1. N 80° E
2. N 62° W
3. S 50° E
4. S 75° W

5. a) 120°   b) S 60° E
6. a) 304°   b) N 56° W
7. a) 253°   b) S 73° W
8. a) 320°   b) N 40° W

Exercise 33e   p. 353

1. a) 12.5 cm, 625 m
   b) 26°, 026°
   c) 52°, 308°
   d) 530 m
2. a) 8.8 cm, 8.8 km
   b) 4.7 cm, 4.7 km
   c) 12.9 km
   d) 066°
   e) 246°

3.

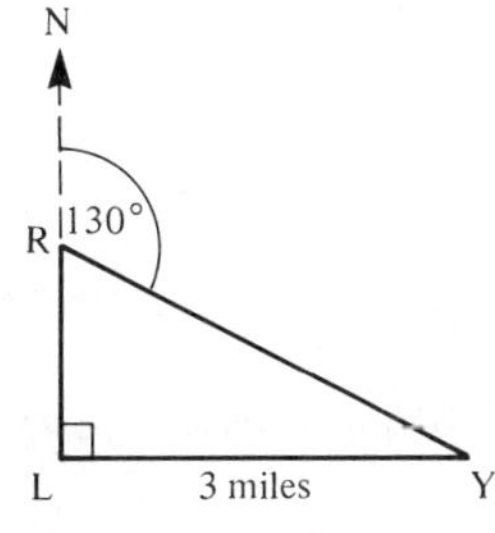

   c) 50°
   d) 90°
   e) 40°
   h) 3.9 miles

4.

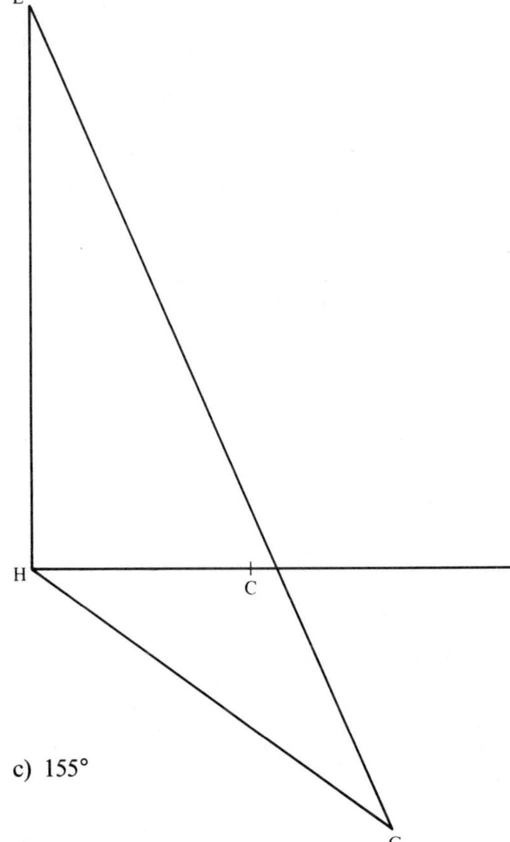

   c) 155°

First Published 1988 by
Stanley Thornes (Publishers) Ltd,
Old Station Drive,
Leckhampton,
CHELTENHAM GL53 0DN

British Library Cataloguing in Publication Data

ST(P) Mathematics 5c
   Answers
   1. Mathematics—For schools
   I. Bostock, L. (Linda),
   510

ISBN 0-85950-930-3

Typeset by Tech-Set, Gateshead, Tyne & Wear
Printed and bound by Ebenezer Baylis & Son Ltd, Worcester